Environmental Humanities: Transformation, Governance, Ethics, Law

Volume 7

Series Editors

Felix Ekardt, Leipzig/Rostock, Germany

Susanne Stoll-Kleemann, Greifswald, Germany

Sustainability, i.e. the demand for long-term and globally practicable lifestyles and economies, is increasingly being understood as the central challenge of our time. But just as today science is often simply equated with natural science, many people think only of the natural sciences when it comes to sustainability science. Natural scientific and technical knowledge of our problematic relationship to nature, resources and climate are undoubtedly important. However, technical change does not happen on its own. In addition, the ecological challenges we face are simply too great and can only be adequately addressed with both behavioural change as well as technological innovation. This is the starting point of this series of publications. Some questions are, for example, the conditions for individual and social change, the means or governance instruments and normative (ethical and legal) issues about the ultimate goals to be pursued. Transdisciplinary approaches should play a special role, i.e. approaches that do not operate from disciplinary boundaries but from questions of content without excessive subordination to established disciplinary dogmas. It is important to the editors that the present series stands for pluralism and expressly gives room to uncomfortable, unexpected and heterodox views and methods. In times in which sustainability research in particular (also) is increasingly influenced by the interests of clients, such openness seems necessary in the interest of finding true knowledge.

Tobias Karcher SJ · Stefan Einsiedel ·
Alberto Irezabal · Theo Haas ·
Amar K. J. R. Nayak · Peter Rožič SJ ·
María Dolores Sánchez Galera ·
Peter Stücheli-Herlach
Editors

Making Hope Possible

On Scaling Up Transdisciplinary Scholarship for Sustainability

Springer

Editors
See next page

ISSN 2524-5708 ISSN 2524-5716 (electronic)
Environmental Humanities: Transformation, Governance, Ethics, Law
ISBN 978-3-032-17718-6 ISBN 978-3-032-17719-3 (eBook)
https://doi.org/10.1007/978-3-032-17719-3

This work was supported by Lassalle-Institut.

This Springer imprint is published by the registered company Springer Nature Switzerland AG
The registered company address is: Gewerbestrasse 11, 6330 Cham, Switzerland

Editors
Tobias Karcher SJ
Lassalle-Institute
Zurich, Switzerland

Alberto Irezabal
Universidad Iberoamericana Ciudad de México
Mexico City, Mexico

Amar K. J. R. Nayak
Professor of Strategy and Chairperson of AIESS Centre
Xavier Institute of Management
XIM University
Bhubaneswar, Odisha, India

María Dolores Sánchez Galera
Senior Research Advisor to the Dicastery for Promoting Integral Human Development
Vatican City, Vatican City

Universidad Carlos III de Madrid
Madrid, Spain

Stefan Einsiedel
Managing Director
Zentrum für Globale Fragen
Hochschule für Philosophie
Munich, Germany

Theo Haas
Lassalle-Institute
Zurich, Switzerland

Peter Rožič SJ
Director of Integral Ecology Research Network
Campion Hall, University of Oxford
Oxford, UK

Peter Stücheli-Herlach
Professor for Organizational Communication and Public Sphere
Zurich University of Applied Sciences
Winterthur, Switzerland

Introduction

By Peter Stücheli-Herlach, Tobias Karcher SJ, Stefan Einsiedel, Theo Haas, Alberto Irezabal, Amar Nayak, Peter Rožič SJ, Maria Dolores Sanchez Galera

We, the editors, believe that humans have to confront their weaknesses, wounds, and the negative consequences of their actions. Otherwise, it will not be possible to live our lives within the constraints of Earth's ecological systems in a manner that ensures our planet's continued existence. This is how the Christian narrative of the expulsion from paradise can be interpreted. This is also the consensus of a modern, "enlightened" humanism that does not rely decisively on theological principles (see the term of the "creature of deficiency" by Gehlen 1950; Pinker 2018: 410-453; Strasser 2020). A good life can only be achieved if humans guard against the dangers they themselves have brought into the world.

This anthropological connection between self-inflicted danger and the need to transcend one's existence (Sloterdijk 2012) forms the backdrop to this publication. In view of anthropogenic, global polycrises, addressing this linkage is as topical and urgent as ever (WEF 2025). Obviously, human-made issues are at the root of the polycrisis. That's both the bad news and the good news: something has to change urgently, and, fortunately, something can change, though it depends on us humans.

Scientific investigation of the crisis thus leads to an imperative for social and political transformation, including, for instance, a substantial decrease of injustice both within societies and between them (Giraud and Sarr 2021). This is particularly important in light of looming food insecurity in many countries of the global south, accelerating climate change, growing difficulties of international and multilateral cooperation, migration issues and the energy crisis especially notable in democratic welfare states in Europe.

While science helps identify and address life-threatening changes on Earth caused by humans (IPCC 2023), it is also imperative that research considers the prospect of a transformation that goes beyond what humans have achieved scientifically so far. What becomes relevant, then, is a perspective that focuses on humans, but goes nevertheless far beyond their current routines and constraints in their institutional, political and everyday action (i.e., their action when "paradise is lost"). And this is precisely where the possibility of hope arises.

The concept of hope initially emerges as a reaction to danger and fear, and even, in some cases, despair and psychological and emotional paralysis. We recognize, on the one hand, that our future on earth is endangered, and on the other, that our means of redemption are not immediately and directly available since they lie beyond our given practical routines and self-perceptions. Nevertheless, the concept of hope simultaneously brings us to a moment of truth regarding our own existence and what could give meaning and justification to it. This truth transcends our seemingly concrete individual personality and our customary habits and routines. Yearning to connect to something greater than our individual self, we realize, "That we hope is precisely because we cannot achieve something on our own" (Grethlein 2024, Introduction; translation by the authors).

These moments of truth thereby connect us as human beings and opportunities for collective social action arise. Hoping, together with others, becomes a priority for human survival and coexistence in our endangered ecological environments. This echoes the spirit of what cultural theorist Raymond Williams once wrote: "It is then to make hope practical rather than despair convincing" (Williams 1989: 209).

The idea that hope can be a response to anthropogenic crisis is not new. It presents decidedly practical prospects and a realistic future on Earth in which humans still play a role. The main reason why the editors have committed themselves to this publication is to make more recent research within particular contexts regarding what these prospects and future could look like more widely known, as well as to develop them further. For example, Macy and Johnstone (2022: 4) argue for what they call "active hope". For them, what matters for the future is how we relate to hope and what we make out of it. They suggest that hope becomes a practice, "A process we can apply to any situation" (ibid.). Indeed, they go on to say: "hope cannot be something we simply 'have'; rather, it is something we do; the term then refers to the practice of moving forward into a future we desire" (ibid.) Based on this, the authors describe a well-tested procedure to practice hope for individuals as well as for groups and communities. And in this regard they agree with Pelluchon (2023), who suggests that hope is fundamentally different from merely waiting optimistically until people with supposedly superhuman talents and abilities will create a better future for us. Hope is therefore a field of practice involving perception, cognition, and reflected action that applies to everyone.

This field of practice is by no means an isolated specialty of those particularly affected or explicitly a specialty of activists. As Krafft (2022) demonstrates, it is a broad and diverse social phenomenon (see also his contribution in this volume).

In summary, we can understand hope as an umbrella term for human practices and individual activities dedicated to transformation in the face of self-inflicted danger. Practices and activities are always configured and carried out in an interconnected material world, and are thus embedded in the here and now and organized along a wide variety of ideas. Hope is, to put it in Schatzki's words, "shared embodied know-how" used to manufacture a future in a way that seems perhaps impossible for those who do not understand that hope is an embodied action (Schatzki 2001: 3).

Thus, for scientific investigation, hope is not only an object. It can also be a frame of reference for scientific practices. However, the conceptualization we follow makes clear that hope isn't at all a specific practice to be investigated exclusively by psychology, theology, ethics, philosophy or ethnography, for instance. It can be investigated–and it has to be investigated–from very different perspectives, and by linking the various approaches and insights of these and other fields together.

Thus, hope could be regarded as a normative prerequisite of what was earlier described as "engaged scholarship" (van de Ven 2007; Barge and Shockley-Zarabac 2008) or of a "transformative science" that has been proclaimed more recently (Schneidewind et al. 2016). The latter denotes a scholarship that is engaged to develop knowledge not only on systems and objectives, but also on pathways of socio-ecological transformation (ibid.). It deals with heterodox paradigms, develops cooperative organizational networks for research activities, and launches initiatives for citizen sciences and open research data as well.

In order to describe how research is understood here, the editors rely on the most recent epistemological and methodological positions that resulted from an intercontinental and intercultural scientific discussion on inter- and transdisciplinarity (Hirsch Hadorn et al. 2008; Rigolot 2020; Merçon 2022; Silva et al. 2022; Vienni Baptista et al. 2023). In principle and in contrast to multidisciplinarity and interdisciplinarity, transdisciplinarity is understood as a process in which different approaches and perspectives do not only complement each other, but merge and thereby develop further together in a dynamic way. The aim is to integrate the perspectives of practitioners as well as of decision-makers systematically across social sectors (Klein 2016: 16; see also Nayak in this volume). This approach can be operationalized among multiple disciplines and professions by deeply exploring common principles, ethics, methods and core design factors that interconnect in any ecosystem. It involves embedded and engaged interactive action research to study and transform reality holistically in its very evolving state and be able to synthesize policies, strategies, and designs of professional as well as of civic action that facilitate flourishing ecosystems. It is considered to have great potential for overcoming the polycrises and achieving sustainable development (Sachs et al. 2019).

However, the enormous challenges involved in implementing transdisciplinarity (see Speranza in this volume) are just as well known as the concept itself. One could say that for the transdisciplinarity of research itself, hope is urgently necessary too. Heterogenous understandings, contexts and everchanging conditions for research both in scientific and policy domains constantly threaten the scope for effective action. Nevertheless, hope is possible insofar as transdisciplinarity is carried out as a practice of exercises in critical self-reflection and dialogue as well as in accepting limitations and failings (Vienni Baptista and Pohl 2023). Such an approach can lead to effective co-production of solutions among participants (Vienni Baptista et al. 2023, Introduction). This requires, however, an openness to embracing diverse ways of learning, investigating and knowing that go beyond traditional constructs to knowledge and models that are developed as they emerge and are routinized in the daily life of citizens, professionals and academics.

Hence, establishing a transdisciplinary scientific practice of hope is a matter of the examination of conditions for human actions, and therefore, the contributions by the humanities and social sciences are indispensable. That is why in this volume, philosophy, theology (see the contribution of Einsiedel&Leibold), information and communication sciences as well as methods such as autoethnography, action research and living experiments play an important role. The same applies to religious, ecclesiastical and spiritual issues, which have been recognized as central to the topic, but are never understood as exclusive or solely valid aspects of a practice of hope.

The contributions to this volume are largely based on presentations given at a transdisciplinary conference entitled "Environmental Justice: Making Hope Possible". The conference was organized in 2024 by the Lassalle Institute in Switzerland, an organization of the Jesuit Order of the Catholic Church, thanks to funding from its own community of executives and private foundations. Single specific invitations to participate also allowed us to round off the content of the volume.

Most of the contributions were first drafted in the form of extended abstracts for the conference mentioned above, but then were revised and expanded for publication in the frame of a second call for papers and a double, friendly peer review process by the editors.

Three questions guided the conference and led to the theoretical and methodological analyses, case studies, autoethnographic documents, and essays presented here:

- How far can "hope" be a legitimate, intercultural, valid and thus "enlightening" concept for international transdisciplinary research and collaboration on ecological transformation and the reduction of injustice? What could it mean regarding science-industry, science-practice, science-policy collaborations?
- How can diverse disciplines (such as economy, ecology, system sciences, agriculture, ethics, policy and management studies, information and communication sciences, sociology and ethnography, practical theology) identify, argue and explain the relevance and impact of "hopeful" perspectives, orientations and practices?
- How could a scientific practice of hope be modeled by theoretical, methodological and empirical means and procedures?

This volume presents a wide range of answers—albeit in an incomplete and inconclusive way.

In the *introductory section* entitled "*The prism of hope*", various conceptualizations of hope as a practice for transdisciplinary research, reflective action and cross-border collaboration are described and documented with examples.

In the section with "*local and regional case studies*", readers will find a well-founded methodological frame for systemic transdisciplinary studies that follows the requirements of the practice of hope, as well as several examples of case studies and action research in the field.

In the final section called the *"autoethnographic space"*, readers can immerse themselves in various contexts of the practice of hope on diverse continents and in institutional contexts. Dense descriptions of the practice of hope are supplemented by autoethnographic testimonials and essays from the field.

Overall, the volume shows conclusively that hope is a feasible, though not easy, transdisciplinary practice in the face of global polycrises. It does not provide any magic formulas for the necessary comprehensive transformation to solve these crises, but it does show concrete opportunities to open a door to what is presumably impossible, but nevertheless necessary, given our current challenging circumstances.

References

Barge JK, Shockley-Zalabak P (2008) Engaged scholarship and the creation of useful organizational knowledge. J Appl Commun Res 36(3):251–265. https://doi.org/10.1080/00909880802172277

Gaël G, Felwine S (2021) L'Économie à venir. Les liens qui libèrent: Paris. ISBN 979-10-209-0979-4

Gehlen A (1940) Der Mensch. Seine Natur und seine Stellung in der Welt (textkritische Edition). Vittorio Klosterman.

Grethlein J (2024) Hoffnung: Eine Geschichte der Zuversicht von Homer bis zum Klimawandel. C.H. Beck. ISBN 9783406821363

Hirsch Hadorn G, Hoffmann-Riem H, Biber-Klemm S, Grossenbacher-Mansuy W, Joye D, Pohl C, Wiesmann U, Zemp E (2008) Handbook of transdisciplinary research. Springer. https://doi.org/10.1007/978-1-4020-6699-3

IPCC Intergovernmental Panel on Climate Change (2023) Climate change 2023: synthesis report. ISBN 978-92-9169-164-7

Klein JT (2010) A taxonomy of interdisciplinarity. In: Frodeman R, Klein JT, Mitcham C (eds) The oxford handbook of interdisciplinarity. Oxford University Press, pp 15–30

Krafft A (2022) Unsere Hoffnungen, unsere Zukunft: Erkenntnisse aus dem Hoffnungsbarometer. Springer. https://doi.org/10.1007/978-3-662-64289-4

Macy J, Johnstone C (2022) Active hope (revised ed.) New World Library: Novato

Merçon J (2022) Investigación transdisciplinaria e investigación-acción participativa en clave decolonial. Utopía y praxis latinoamericana 27(98)

Pelluchon C (2023) L'espérance ou la traversée de l'impossible. Rivage, Paris

Pinker S (2018) Enlightment now: the case for reason, science, humanism, and progress. Penguin Random

Rigolot C (2020) Transdisciplinarity as a discipline and a way of being: complementarities and creative tensions. Humanit Soc Sci Commun 7:100. https://doi.org/10.1057/s41599-020-00598-5

Sachs JD, Schmidt-Traub G, Mazzucato M, Messner D, Nakicenovic N, Rockström J (2019) Six transformations to achieve the sustainable development goals. Nat Sustain. https://doi.org/10.1038/s41893-019-0352-9

Schatzki T (2002) Introduction: practice theory. In: Schatzki T, Knorr-Cetina K, von Savigny E (eds) The practice turn in contemporary theory. Routledge, pp 1–14.

Schneidewind U, Singer-Brodowski M, Augenstein K (2016) Transformative science for sustainability transitions. In: Brauch H, Oswald Spring U, Grin J, Scheffran J (eds) Handbook on sustainability transition and sustainable peace. Hexagon series on human and environmental security and peace, vol 10. Springer. https://doi.org/10.1007/978-3-319-43884-9_5

Silva F, Coward F, Davies K, Elliott S, Jenkins E, Newton AC, Riris P, Vander Linden M, Bates J, Cantarello E, Contreras DA, Crabtree SA, Crema ER, Edwards M, Filatova T, Fitzhugh B, Fluck H, Freeman J, Klein Goldewijk K, Williams A (2022) Developing transdisciplinary approaches to sustainability challenges: the need to model socio-environmental systems in the Longue Durée. Sustainability 14(16). https://doi.org/10.3390/su141610234

Sloterdijk P (2012) Du musst Dein Leben ändern: Über Anthropotechnik. Suhrkamp

Strasser P (2020) Die Sprengkraft des Humanismus: Ein Beitrag zur Politik der Seele. Karl Auer

Van de Ven AH (2007) Engaged scholarship: a guide for organizational and social research. Oxford Academic. https://doi.org/10.1093/oso/9780199226290.001.0001

Vienni Baptista B, Pohl CE (2023) Exploring interdisciplinarity and transdisciplinarity as knowledge regimes: a heuristic tool for disentangling understandings in academia and policy. Sci Technol Hum Values 49(6):1309–1348. https://doi.org/10.1177/01622439231216789

Vienni Baptista B, Fletcher I, Lyall C (eds) (2023) Foundations of interdisciplinary research: a reader. Bristol University Press. ISBN 978-1529225747

WEF World Economic Forum (2025) Global Risk Report. WEF

Williams R (1989) Resources of hope: culture, democracy, socialism (ed. by Robin Gable). Verso

Contents

Part I
The Prism of Hope: Conceptualization of Hope for Transdisciplinary Research and Cross-Border Collaboration

Chapter 1
Reflections on Transdisciplinary Research for Sustainable Food Systems

Chinwe Ifejika Speranza, Stellah Mikalitsa Mukhovi, and Boniface Kiteme

Abstract Reflecting on transdisciplinary research can enhance clarity on its potential to promote scientific innovation and contribute solutions to complex societal challenges. We analyse a transdisciplinary project on food systems and the applications of its resulting tool, the Food Sustainability Assessment Framework - "FoodSAF". We employ the general phases of transdisciplinary research, encompassing joint problem framing, co-design, collaborative research, results validation, piloting negotiated transformative action(s), and joint evaluation of research impact. We discuss lessons, challenges, opportunities, and strategies of operationalizing such transdisciplinary research, illustrating with concrete examples from the project. Insights show that the research process depended on the researchers' capacities, their freedom to research, their willingness to break disciplinary boundaries and social differences, to co-create knowledge by learning new methods and acquiring collaboration skills. The research also depended on the conduciveness and supportiveness of the research contexts, stakeholder willingness to collaborate, and the availability of financial and time resources for the research. The transdisciplinary research contributed to positive outcomes for the involved stakeholders. While the focus on one research project does not cover the breadth of research project types and experiences, this contribution provides an opportunity to reflect on research processes and their potential to become more collaborative and transformative.

C. Ifejika Speranza (✉)
Institute of Geography, University of Bern, Bern, Switzerland
e-mail: chinwe.ifejika.speranza@unibe.ch

S. M. Mukhovi
Department of Geography, Population, and Environmental Studies, University of Nairobi, Nairobi, Kenya
e-mail: smmukhovi@uonbi.ac.ke

B. Kiteme
Centre for Training and Integrated Research in Arid and Semi-Arid Land Development, Nanyuki, Kenya
e-mail: b.kiteme@africaonline.co.ke

T. Karcher SJ et al. (eds.), *Making Hope Possible*, Environmental Humanities: Transformation, Governance, Ethics, Law 7,
https://doi.org/10.1007/978-3-032-17719-3_1

Keywords Transdisciplinary research · Co-production of knowledge · Sustainable food systems · Stakeholder collaboration and social learning

1.1 Introduction

Persisting crises such as land degradation, biodiversity loss, climate change and food insecurity have shifted expectations about how knowledge must be produced and used to achieve scientifically informed social change (Intergovernmental Science-Policy Platform on Biodiversity and Ecosystem Services (IPBES) 2024). The multidimensional nature of these crises has led to the realization that approaches that integrate different disciplinary expertise and knowledge systems are necessary to diagnose them and identify solution options (Norström et al. 2020; Oberlack et al. 2019). Further, the need and hope to address these crises more effectively and to enhance the relevance of science for society has led to proposals of a transdisciplinary research (TDR) approach.

The rationale underpinning TDR is that by engaging stakeholders as co-producers of knowledge, more socially relevant and robust research outcomes are likely to be achieved (Kiatkoski Kim et al. 2022). TDR holistically integrates knowledge from different disciplines (interdisciplinary) with knowledge and perspectives from non-academic actors throughout a research process (Karrasch et al. 2022; Ustundag 2022). Therefore, it fosters meaningful collaboration and dialogue across stakeholder groups, enabling the co-production of knowledge to address societal and scientific challenges while simultaneously advancing knowledge and social transformation (Ustundag 2022). Various principles, concepts, and frameworks have thus been proposed to guide TDR (Pohl et al. 2021). While this call to transform research approaches and methods is important, the questions of how and to what extent TDR is actionable are considerably debated (House et al. 2024; Rigolot 2020). Among others, empirical contexts often challenge the intent of a TDR approach (Hoffmann et al. 2017; Kläy et al. 2015; Schneider et al. 2022). Thus, the expectations accorded TDR of achieving outcomes that are not only beneficial to science in terms of advancing knowledge but also to society in terms of actionable knowledge to solve societal problems require continuous reality checks.

In this contribution, we reflect on and derive lessons from TDR practice based on a recently concluded transdisciplinary research project: the "Towards food sustainability: Reshaping the coexistence of different food systems in South America and Africa" (Jacobi et al. 2020; Rist et al. 2016). In what follows, we describe the research project, present an analytical framework that enables reflection and discussion of the research processes, discuss the lessons we have learned and then conclude.

1.2 Case Study Research Project

The "Towards Food Sustainability" research project (Table 1.1) (henceforth research project) aimed to analyse factors influencing food sustainability, to identify and promote strategies and policy options to enhance the sustainability of different coexisting food systems at multiple levels in Bolivia and Kenya (Jacobi et al. 2020; Rist et al. 2016). One of the co-created outputs of the project is the Food Sustainability Assessment Framework (FoodSAF). The FoodSAF is a participatory food systems assessment tool co-created by combining research outputs from all the studies in the two case-study countries of Kenya and Bolivia to identify the indicators that can best explain food sustainability and was validated in non-research countries—Zambia, Ghana, Brazil, Colombia, during its second phase and in a further application of the FoodSAF in Nigeria. Elaborations in this contribution focus only on the Kenyan cases. Besides the principal investigator, the research team comprised other co-investigators from different organisations in Switzerland and in the case study countries, a project coordinator, various co-supervised PhD researchers, Master and Bachelor students.

The project was funded by the Research for Development (r4d) programme of the Swiss Agency for Development and Cooperation and the Swiss National Science Foundation promoting research partnerships between Switzerland and various countries in Africa, Asia, and Latin America. This funding partnership aimed to provide

Table 1.1 A descriptive profile of the "Towards Food Sustainability" research project

Project variables	Project description
Project title	Towards food sustainability: reshaping the coexistence of different food systems in South America and Africa
Research goal	Enhance knowledge on how to improve the sustainability of food systems
Research approach	Transdisciplinary; interdisciplinary, disciplinary
Current research stage	Concluded
Pre-funding/seed funding	Yes
Collaborative research project or career funding	Collaborative research project
Funding duration	Six years (2 phases of 3 years) plus two for implementation (01.2015 –12.2017 and 01.2018 - 08.2021) Some research activities in Kenya took longer due to COVID-19 pandemic and a severe drought. Project activities were suspended to allow the farmers to cope with drought and COVID-19
Transformative pilot actions (TPA)	A TPA is a short-term joint science-practice action that implements and tests innovative research findings to solve societal problems. Planned in the proposal, funded and implemented
Joint evaluation of research impact	Science-driven

knowledge that is relevant for both science and societal development to policymakers (Meyer and Wehrli 2024). Besides a food systems approach, the project adopted a TDR approach from the outset.

1.3 Methodology

We adopted a reflexive research approach to examine the process of knowledge generation in the TDR project. To structure the reflections, an adapted version of the TDR framework proposed by (Pohl et al. 2017, 2021) was applied (Fig. 1.1).

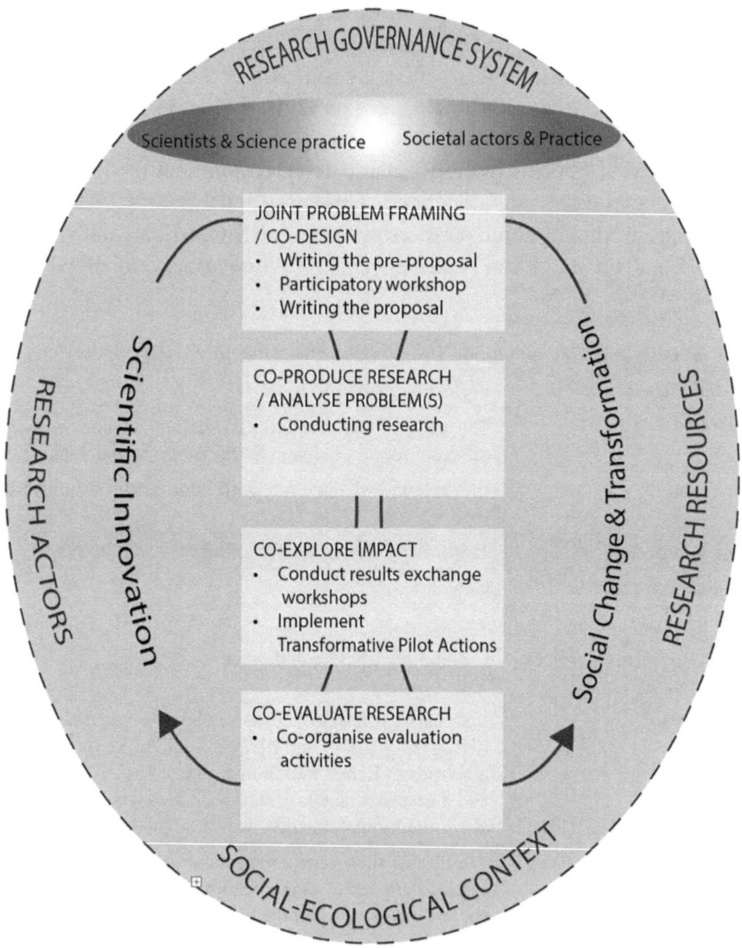

Fig. 1.1 The processes of TDR and associated framing conditions (Own illustration with adaptations from Pohl et al. 2021)

The framework (Fig. 1.1) depicts TDR, and participatory research approaches, as embedded in a context of research (focal) action situations (cf. Ostrom 2011) or moments of interactions (boxes) among the involved researchers and societal actors, the resources at their disposal to conduct the research and the research governance system. Research thus occurs within a context that can influence the research itself but is also influenced by the research (Fig. 1.1).

At the top of the ellipse (Fig. 1.1), we see the research governance systems such as research funders and their institutions (rules and norms) as well as the institutional context in case study countries regarding research. These together define eligibility criteria in a funding scheme and set the rules for researchers or other social actors to participate in the action situation (McGinnis and Ostrom 2014; Ostrom 2011) of framing the research problem (co-defining/co-writing the proposals), analysing the problem (conducting research) and exploring impact (communication, dissemination, results exchange, pilot actions, research evaluation). Hence such institutions provide the supportive or constraining basis for researchers and other social actors to participate in a research process. Further, a researcher and other involved actors (Fig. 1.1, left side of the ellipse) have various competencies that qualify them to engage in the research process. Sometimes, establishing such a research consortium with relevant science and non-science actors takes time and must be considered in the research process. Researchers may also be able to tap into resource systems (Fig. 1.1, right side of the ellipse, research resources) set aside by their research organisations such as pilot/seed funding schemes to enable them to prepare a research proposal for funding and to engage with societal actors and practitioners prior to preparing a larger (transdisciplinary) research project.

The column within the ellipse illustrates a TDR process with the coming together of actors and knowledge from the realm of science that seeks to identify facts (left side of the column) and non-academic actors from the realm of societal practice that seek to find solutions to societal problems (right side of the column) (Pohl et al. 2021). These knowledge communities (e.g. academic and non-academic actors active in the research activities) come together at instances (lines from the realm of science and of societal practice) to (1) frame the research problem (writing the preproposal, conducting participatory workshops, and writing the full proposal), (2) analyse the identified/prioritized problem (conducting the research), (3) explore research impact for better understanding sustainability issues and acting on them (Pohl et al. 2021). This includes communicating and disseminating research results, conducting results exchange workshops, Transformative Pilot Actions (TPA), and lastly, evaluating research (Fig. 1.1). Being a process, the TDR stages can be explained also as *focal action situations* with interactions (McGinnis and Ostrom 2014) and outcomes at every iterative stage (Fig. 1.1). Continuing with this analogy of a research process as a concatenation of focal action situations, we can identify such critical moments of interaction among actors involved in the research process as depicted in Fig. 1.1 (middle column).

Thus, research as a social process means managing stakeholders' and researchers' expectations, balancing descensus with consensus in the spectrum from disciplinary to TDR in a broader social-ecological context (Fig. 1.1, bottom of the ellipse).

Drawing on Fig. 1.1, the reflections on the research process are structured as follows: (1) Research context, (2) Joint framing of the research problem, (3) Collaborative problem analysis and knowledge co-production, (4) Exploring societal and scientific impact and paths to transformation, and (5) Joint evaluation of research impact.

1.4 Reflecting on the Research Context

As shown in Fig. 1.1, the research context can be conceptualised as an interaction of four components–(1) research governance system(s), (2) the research resources and (3) the research actors comprising the academic and non-academic actors and (4) their interactions to shape or be shaped by the social-ecological context.

Research governance context encompasses the socio-political norms, that is, the formal and informal rules to adhere to when engaging with research actors. For the research project, agreements were signed among the partners (e.g. international and local project proponents) on the research procedure, roles, responsibilities and remunerations. Research permits had to be obtained from the relevant national authorities (e.g. the Kenya National Commission for Science, Technology and Innovation) with the support of the local research partners and official letters from the research organisation and/or the local research partners (e.g. the Kenyan Centre for Training and Integrated Research in Arid and Semi-Arid Lands Development—CETRAD). Informally, oral agreements were made with local communities via their local authorities (e.g. County Administration levels, Chiefs, Elders) and informing the local government authority and obtaining, through courtesy visits, their informal or formal written consent to research being conducted in their areas of jurisdiction. Similarly, consent was obtained from the research participants (e.g. the food systems' actors) after explaining the research project to them. Compliance with these written and unwritten rules, including research ethics of the involved universities, set the stage for further research activities.

The project received funding from the Research for Development (r4d) programme, of the Swiss Agency for Development and Cooperation and the Swiss National Science Foundation with most research occurring in countries in the Global South. The involved academic and non-academic actors in the project were thus dependent on this funding even though they also made in-kind contributions. This created a financial power imbalance between researchers from Switzerland (Global North) and those from developing and middle-income countries (Global South). In the second phase and in the TPAs (last three years), the Kenya national and County governments, and individuals (including farmers) made considerable financial and in-kind contributions to co-implement the TPAs. Related to funding is the time allocated to research activities by academic and non-academic partners. Fieldwork was mainly by early career researchers in their bid to gain qualifications in the research field. Senior researchers were mainly supervisors, and their busy schedules of teaching and other tasks constrained their engagement in fieldwork to mostly the initial phases and critical moments, such as the initial use of research instruments. In the project, financial provisions existed for local senior researchers to engage and thus they had the obligation to conduct the research activities and were directly responsible to the funder. Although not deeply affecting the research quality of the project, funding and their allocation influenced the extent to which academic and non-academic actors could engage in the research process. This was, for example, through tailoring the number of workshops or number of participants to available funding.

Local actor perceptions of the abovementioned financial capacities affected the research context. The colonial legacy still permeated the relationships between European researchers, African researchers working in Europe and African local researchers in the project—in some cases, respondents and other societal actors perceived those working in Europe as having the financial resources to solve problems and cover costs.

The characteristics of the *research actors* (e.g. gender, age, language, wealth, literacy levels) are important aspects of the research context (Fig. 1.1). For instance, access to and ability to exchange knowledge depended on the literacy level (ability to read and write) of involved community actors, which affected the research project. Many survey respondents, Focus Group Discussions (FGD) or workshop participants completed 6–7 years of primary education, while a few participants did not attend school. The national language, often from the colonial period (English) was not the preferred means of communication especially in rural areas where people were more comfortable and could communicate well in their local languages (e.g. Swahili, Kikuyu, Kikamba). Research methods were thus adjusted to fit the literacy context. For example, methods such as FGD and participatory workshops were applied. Where household surveys were conducted, trained local field research assistants (often University Bachelor graduates or Master students), who spoke the local language, were engaged and did simultaneous translation as they administered the questionnaires to the households or documented the workshops and FGDs.

Availability and access to resources and study sites influenced the research process and outcomes (Fig. 1.1). Resources can be in terms of explicit funding for a pilot process before other TDR activities (e.g. funding to organise a participatory workshop in preparation of the full proposal), availability and access to local actors and infrastructure, once in fieldwork. Access to study sites also depended on their proximity to urban areas. Accessing rural areas, especially remote rural areas needed the arrangement of dedicated transport (e.g. hired taxis, and motorcycles) and communication infrastructure (e.g. mobile phone connections to multiple communication networks), and even with a dedicated communication infrastructure, poor communication networks meant that research teams were cut-off from communication for long hours and, in some cases, days. In some cases, access to reliable infrastructure was easy as we could rent vehicles of partner organisations without having to make the arrangements ourselves. Getting to study sites, be it moving from one household to another household during a household survey, required trekking for relatively long distances and avoiding risky situations.

Such risky situations concerned environmental risks and social tensions and conflicts as may be experienced in different social-ecological contexts (Fig. 1.1). Through local partner knowledge, such areas were avoided and risks of water-borne diseases, snake bites and scorpion stings when in agricultural fields, managed. Prior risk management plans also helped to prepare strategies for any unforeseen circumstances, but in a few cases misunderstandings or small conflicts interrupted the field research activities. In some cases, potential respondents requested payment for being interviewed but the research project had no provisions for such payments hence such interviews were not conducted. In other cases, some researchers, such as those working on legal documents did not see the value of fieldwork because they preferred to analyse gaps in policy and legal frameworks through desk review. However, they acknowledged it was important to confirm how such frameworks were applied at practice level, and whether the farmers had knowledge of their existence.

In the study sites, political situations influenced research activities, whether the research results could be implemented or not and how. In the project, implementing research findings into transformative actions in Kenya was successful, in part due to its fit with government programmes to improve agriculture and food security, thereby motivating government authorities to engage. Thus, socio-economic-political processes influenced the success or failure of transformative actions arising from the research findings.

Considering the above-mentioned contextual factors, pre-existing partnerships were crucial. A pre-existing partnership enables an upcoming research project to build on established trust relationships hence reducing the effort needed to build new partnerships and trust. The project grew out of existing research and development partnerships. The local research partners were officially proponents of the research project and thus directly received funding from the funding scheme and were accountable to the funder through the project coordination for the tasks they assigned themselves. The research project was thus dependent on partnerships with local scientists who not only facilitated access to research sites but also conducted a considerable part of the research themselves.

1.5 Reflecting on the Research Process

According to Fig. 1.1, the research process can be categorised into four main steps—joint problem framing/co-design, co-producing research/analysing problems, co-exploring and co-evaluating research impact. The following sub-sections are thus structured according to reflections on these iterative steps and further detailed in Fig. 1.2.

1.5.1 Joint Framing of the Research Problem

Initiating a TDR project in a specific context requires discussions among the societal actors including researchers. Figure 1.2 illustrates the detailed research steps.

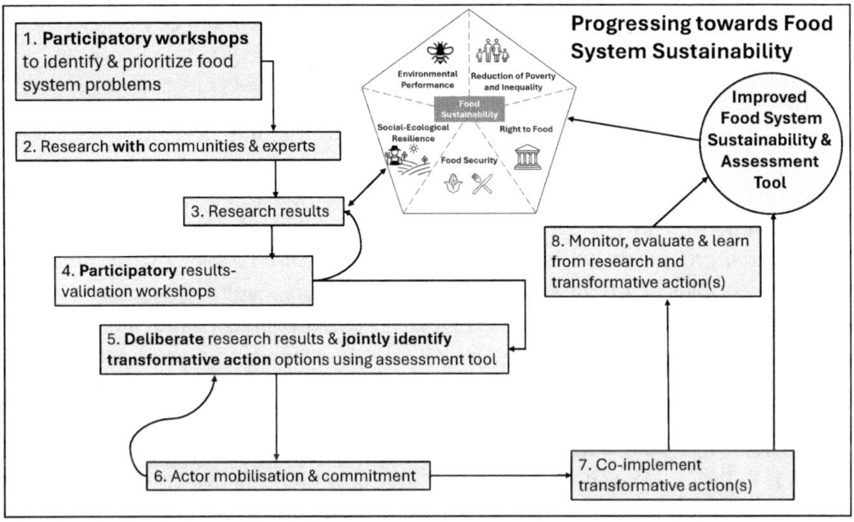

Fig. 1.2 An illustration of the transdisciplinary research activities using the example of the Food Systems Sustainability project

Co-designing the research goals was centred on the problem of food insecurity in Kenya, which has been well known to local and foreign researchers. The TDR project was locally coordinated by CETRAD, a bilateral research organisation funded by both the Swiss and Kenyan governments. Senior researchers of CETRAD together with senior researchers at the Universities of Bern and Geneva, Switzerland, and the Universidad Mayor de San Simón, Cochabamba, Bolivia, developed the pre-proposal based on which funding was obtained for a stakeholder workshop to discuss and set the research priorities (Fig. 1.2, step 1).

A researcher from Bern travelled to Kenya and together with CETRAD colleagues, government officers, NGO officers and community representatives, organised a scoping workshop where the research topic priorities were set after discussing the challenges facing societal actors in the Mount Kenya region. The workshop at the beginning also helped to finetune the methods to identify other relevant stakeholders. A further joint fieldwork in the Mount Kenya region, involving both academic and a wider group of societal actors, contributed to updating knowledge about the study context. Insights from the scoping workshop and field visit were fed into developing the main project proposal, which subsequently got funded. Thus, coming together to frame the research problem was already before and after the pre-proposal development as well as before the full proposal development and subsequent project funding, and laid the foundation for subsequent research activities.

1.5.2 Collaborative Problem Analysis and Knowledge Co-production

After framing the research problem including prioritizing key topics and focus geographical areas, collaborative problem analysis included collaborative writing of the full proposal by the project consortium and meetings until the full proposal was submitted. With project approval and inception workshops, each researcher and research group concentrated on their specific research questions. Besides other research methods, participatory methods (e.g. Focus Group Discussions—FGD and participatory workshops) were applied with the non-academic actors actively involved in co-producing knowledge. These non-academic actors sometimes co-steered the research process (Fig. 1.2, steps 2–3), such as through collecting and curating data or moderating FGDs and workshops.

The research topics influenced knowledge co-production. The collaborative research problem centred on addressing access to and control of resources such as land, and water. This was a sensitive issue socially and politically during the time of the research. Major issues are related to land tenure, gender disparities, low farm-gate prices, water abstraction by upstream commercial land users that often reduced water availability for downstream users (mainly pastoralists and agropastoralists), pasture conditions and access as well as drought. Farm management strategies exerted some externalities such as spraying of agrochemicals and their non-desirable drift to neighbouring croplands where farmers were practising agroecology. These problems influenced the choice of transformative actions to co-implement as feasibility was a key factor for the research participants to prioritize an action point.

The TDR project applied methods from the social- and natural sciences and integrated methods depending on the specific research questions, to enable knowledge co-production. Local research assistants were engaged and trained to support data collection in addition to the local researchers who were themselves doctorate or post-doctorate researchers in the project. Common data collection methods included administering questionnaires to households, conducting various expert interviews, key informant interviews, participant observations, gender-differentiated FGD, life cycle assessment, and participatory workshops. Soil and water samples were also analysed. Insights from different disciplinary perspectives were thus integrated to achieve research goals, while contributing to disciplinary insights.

Thereby the TDR was also affected by the quality of social interactions between the various societal actors involved (Larson et al. 2020). Openness for interaction varied across case study sites with some actors open to interaction while others were unwilling to engage in the participatory processes. Schneider et al. (2022) argue that the willingness to interact depends on the benefits actors expect to gain. Thus, building trust was key. Further, pre-existing partnerships of local researchers and their organisations with societal actors and the long-term partnerships contributed to trust-building.

Stakeholder engagement in TDR and participatory research were equally key. In some locations, some societal actors expressed research fatigue due to previous research not having significantly contributed to improving their living conditions. In other cases, even though (expected) benefits were less tangible, societal actors still engaged as some of them found the topic of addressing food insecurity beneficial in a non-monetary sense. Yet stakeholder engagement depends on the quality of communication among research participants.

Such communication depends on mutual language and understanding (Rist et al. 2006). Language barriers between researchers and other societal actors, especially in the rural areas, meant that interpreters had to be engaged and trained not only to understand the purpose of the research activities but also the intricate terminologies. Scientific concepts needed to be translated in a way that made them understandable to all societal actors. In the project, terms like resilience, environmental performance, and ecosystem services were used. Local actors conversant in the local language and culture as well as conversant or trained in the use of scientific concepts, referred to as bridging actors, were engaged. These bridging actors thus served as knowledge exchange agents but also as filters as they translated between the researchers and other societal actors, and misunderstandings sometimes occurred.

Sharing knowledge is thus critical for knowledge co-production. Knowledge held by societal actors is also not always accessible and depends on the actor's willingness to participate in the research process. As is the case with good scientific practice, acquired knowledge from societal actors needs to be anonymised, and reflections must be made on how to share sensitive data and knowledge (e.g., household livelihood data). Sharing knowledge also raises the question of the role of scientists and researchers, whether their role ends with generating knowledge and showing how the new knowledge can be implemented. This includes questions of whether they need to go further to advocate for and implement actions to achieve social change, no matter how small, or whether they must become activists to advocate for the consideration of knowledge in social action. With the core purpose of TDR being to contribute to solving complex societal problems, through integrating knowledge from academic and non-academic realms, researchers assume new roles as knowledge mediators for which they might not have prior training.

The level of knowledge exchange also depended on the social structures, which were often hierarchical. This meant that in a group setting, such as in workshops, socially higher-ranked actors (e.g. elders) often had the word, and other social actors, including women, rather listened than voiced their opinions. Professors, senior researchers, government officers and traditional rulers often voiced or were expected to first voice their opinions while other actors with less social power were often reluctant to express their views. Traditional (and conventional) norms and roles thus made it difficult for all societal actors present in a workshop to voice their experiences and knowledge to comparable degrees. However, through gender-differentiated focus group discussions and household surveys, men and women groups as well as less socially powerful and less wealthy actors were provided a platform to express their views and discuss options to arrive at joint decisions.

Further, a North–South knowledge hierarchy was also perceived, with some local actors viewing the Europe-based researchers as having "better knowledge" and greater capacity to address local problems compared to their locally based counterparts. Hence, local actors may, in effect, devalue their own expertise and abilities to propose solution options that may be workable and contextually appropriate. Thus, during such co-production exercises, it became imperative to adopt strategies that empower local social actors and enable their active contributions to creating solutions relevant to their contexts. Unexpressed tensions also arise as researchers from the Global-North or from outside the research context might be time-bound due to their academic calendar while societal actors in the study sites had to adjust their schedules, livelihood strategies and immediate priorities to accommodate researcher requests. Exploring the boundaries and extent to which both contextual knowledge and external forms of knowledge can be bridged in knowledge co-production was thus important.

1.5.3 Exploring Societal and Scientific Impact and Paths to Transformation

Exploring societal and scientific impact of TDR can be *throughout the research process since the research process is iterative and non-academic actors are expected to be engaged throughout*. In the project, this exploration was mainly towards the end of the research activities when most research results had been realised (Fig. 1.2, steps 3–7).

Many results of the project revealed obstacles to the sustainability of focus food systems. The project thus led to the development of a Food Sustainability Assessment Framework (FoodSAF) tool, a synthesis product that assesses the sustainability of food systems (Rist et al. 2021) and identifies solution options in collaboration with societal actors. In subsequent results exchange workshops with societal actors (Fig. 1.2, steps 4–7), project findings were presented and the FoodSAF tool applied to different case studies.

Results from research with the dairy farmers in Umande, Mount Kenya region, showed that farmers obtained only farm-gate prices from milk sales as they depended on traders and middlemen to market their milk hence only derived minimal benefits (Llanque et al. 2021; Mukhovi et al. 2024). Farmers had no option to preserve their milk adequately hence were under pressure to sell at low prices. To improve their benefits from the value chain, based on research findings and dialogue with farmers in workshops guided by the FoodSAF Tool, farmers chose to engage in milk cooling and bulking instead of selling unprocessed milk as a path towards improving their incomes. Thus, the project's result exchange workshops led to Transformative Pilot Actions (TPA). A TPA, is a short-term joint science-practice action that implements and tests innovative research findings to solve societal problems (e.g. train farmers, enhance farmer position in value chain through milk processing). This would be expected to lead to a major positive change or having the potential to do so, thereby validating the research findings.

After identifying the potential TPA with the Umande farmers, it turned out that, not being organised in groups, the individual farmers could not improve their bargaining power and afford to obtain training on improved dairy farming practices. The Umande TPA thus comprised creating the Umande Farmers' Cooperative Society, training cooperative members on best practices in dairy farming by extension staff from the Ministry of Agriculture, Livestock and Fisheries, and by the Cooperatives department. The TPA also involved the construction of a cooler house and the installation of a milk cooler. Through the Umande Farmers' Cooperative Society, created during the FoodSAF application, farmers could combine their financial resources and have a unified voice towards government authorities and other societal actors. Support was provided in various forms to the cooperative, such as through meetings organised by CETRAD to disseminate research results and mobilize societal actors; through institutional and financial support from CETRAD and the research project in the scheme of the r4D funding, from local leaders (Member of Parliament and Member of County Assembly), the County government, and the national government. The Kenyan national government via the Ministry of Agriculture, Livestock and Fisheries, Directorate of Agriculture and through the County Department of Cooperatives provided a 3,000 L milk cooler. The Laikipia County government provided land to the farmers' cooperative to build a house for the milk cooling plant and support to establish a milk distribution network, while the cooperative members contributed labour and materials (Mukhovi et al. 2024).

These innovative pathways enabled dairy farmers to influence milk prices and avoid selling their milk at low prices. According to the farmers, they could increase their milk yields and prices as well as invest in their agricultural production, for example through processing milk into yoghurt, thereby increasing job opportunities and improving their food security (Bessa et al. 2020). Farmer training as a component of the project, contributed to improving labour and farm productivity and efficient use of agrochemicals. Learning through exchange visits with farmers who belong to established milk processing facilities like the Meru Farmers' Cooperative Society increased farmer confidence in their own knowledge. Further, farmers reduced the cost of milk processing by leveraging economies of scale. The training on post-harvest management contributed to increase maize yields from about two to three 90 kg bags in conventional farming to eight to twelve 90 kg bags of maize per acre through the adoption of conservation agriculture practices proposed by FoodSAF and related research (Bessa et al. 2020).

In another site in Kenya, the FoodSAF tool was applied to assess the sustainability of the Makueni agro-pastoral food system resulting into the Makueni County TPA. In this case, the TDR, which comprised mainly participatory workshops, benefited from prior long-term research and development partnerships. The Makueni FoodSAF TPA could thus build on knowledge from long-term development research initiatives of the CETRAD in the Makueni County on conservation agriculture and other previous research (Ifejika Speranza and Kiteme 2020). Applying the FoodSAF tool to the Makueni case in workshops with societal actors revealed a low level of social self-organisation among the farmers. Through deliberations and ranking solution options, farmers therefore chose to enhance their social self-organisation by forming the

Makwambu Community Based Organization (CBO) and improved their positions in the food value chain. Through a similar process of mobilization by CETRAD, co-funding from the project and the farmers, land was acquired to build a grain aggregation storage house. The grain storage house was to enhance farmers' positions in the grain and pulses value chain (e.g. pigeon peas, green grams, beans, maize) and forestall sales immediately after harvest, which hitherto exposed farmers to low market prices (Ifejika Speranza et al. 2008). It enabled cooperative members to aggregate their grain and pulses for storage, to reduce the influence of middlemen and improve their flexibility to sell their grains when market prices were profitable. Such a strategy increased their incomes and improved their livelihoods. Self-organisation into a CBO also strengthened their voice towards government authorities and other societal actors.

Besides societal impact and knowledge generation, the "Transforming food systems" project had various scientific effects: It expanded partnerships and contributed to staff development in the Global South and Global North—the PhD alumni and postdoctoral researchers progressed in their career to various positions, as senior researchers, university lecturers, programme directors or professors. The project produced various scientific publications, videos, user-inspired outputs and other materials. It was a finalist in an award competition on "Transformative Agroecological Research for Development". The project consortium organised various academic and knowledge transfer events and project members presented the research results in various platforms. The project enabled learning and practicing TDR methods and produced a tool that is used to assess the sustainability of food systems in various world regions.

Finally, it is important to highlight that exploring societal and scientific impact of TDR and other research took time–from the scoping workshops to project conclusion, it took about six years, and the subsequent result exchange, and dissemination period an additional two years. The process of prioritizing, implementing and evaluating transformative actions also took time. Further, the TPAs were also affected by the COVID-19 pandemic and a drought in Laikipia- (Umande) and Makueni Counties. However, the benefits of such a TDR approach are not just in the TPAs at the end of the research process but also the inclusive research process, which was not always easy. The inclusive and participatory process enabled social learning by all involved actors through increasing awareness about the sustainability challenges of the focus food systems and exchanging knowledge on how to address them.

1.5.4 Joint Evaluation of Research Impact

In TDR, a joint evaluation of research impact is ideally expected. However, joint evaluation was not explicitly and formally conducted in the project. Rather, evaluation of research impact was largely conducted by the researchers themselves, through reporting to the funder their research outputs such as publications, the developed tools such as the FoodSAF Tool (Rist et al. 2021), ex-poste video documentaries of the research outcomes by interviewing farmers and other involved actors (Bessa et al. 2020; Ifejika Speranza and Kiteme 2020) and informal exchanges on the status of the implemented TPAs as well as follow-up projects based on the FoodSAF (Makri and Patil 2021). The qualitative research impact evaluation focused on the socio-economic transformation from the perspectives of local actors and researchers, e.g. how the milk cooling plant improved farmers' position in the value chain, on the qualitative perceptions of improved wellbeing by the farmers through the video documentaries. From the perspective of researchers, impact was also through building capacities at the universities. Still, because the project was embedded in the Swiss r4d Programme, research impact was evaluated in the context of the r4d programme, which showed that in many respects, the r4d programme was innovative and ahead of its time (Meyer and Wehrli 2024).

1.6 Discussion

The elaborations in this paper show that research context, as well as the researchers' capacities, influence the possibilities of TDR. Contexts can be challenging for all study sites. The "Towards Food Sustainability" project was defined as a TDR project and provided *funding* to implement transdisciplinary processes (continuous collaboration and co-implementing research results into transformative actions). Elements in the project can be understood as forms of action research, which involves the engagement of affected persons (stakeholders) in identifying, analysing and acting on a research problem (Lewin 1946; Snapp et al. 2023). Hence, inclusive approaches are important to ensure that all relevant actors are involved in knowledge co-production and marginalised actors have a voice (Schneider et al. 2022).

Co-producing new knowledge in a North–South context depends on many factors in multi-site contexts, among them, functioning research institutions, discourses about science-policy-society relations (e.g. engagement with relevant stakeholders at all levels, from local-national policy levels), research resources and infrastructure, social dialogue cultures, power disparities and knowledge hierarchies, literacy and ability to communicate, and perceived societal values of research (Norström et al. 2020; Schneider et al. 2022). As the site manifestations of these factors might vary, their fit across the multiple research sites (Global North, Global South-national-local levels) is critical for a TDR to achieve its set goals.

Field research whether TDR, interdisciplinary, disciplinary or participatory can be challenging hence requiring adequate knowledge of field research contexts and adaptive planning (Adler et al. 2018; Ott and Kiteme 2016) as well as planning for delays in the research process. Skills in intercultural communication are critical to ensure that researchers adequately consider language barriers, differing cultures and dialogue cultures (Schneider et al. 2022).

Besides context-factors, social relational factors also played an important role. Long-term partnerships (Lux et al. 2019; Wiesmann et al. 2011) and boundary organisations—organisations interfacing science and policy (Gustafsson and Lidskog

2018), such as local research and development organisations (e.g. CETRAD or even local NGOs) or through real-world labs helped build the *trust* needed to enable open dialogue and become responsive to local contexts and needs (Bergmann et al. 2021).

Further, the elaborations show that research policies influence research practice, whether TDR or non-TDR, so also research realities, which might force a researcher to adapt the research process. Adaptation may be in terms of increasing involvement/engagement by practitioners and societal actors, but this depends on the flexibility that the researcher has, in terms of rules, norms, resources and time. As with other research approaches, TDR is exposed to political and governance contexts, which can affect all stages/phases of a research, from stakeholder inclusion to defining the research problem/questions, implementing TPAs that are informed by the research results, research uptake and impact (House et al. 2024; Pärli 2023). Exploring pathways to transformation thus requires context specificity in terms of considerations of the site-specific values, norms, agency, local aspirations, as well as governmental policy directions while considering power relations, amongst others (House et al. 2024).

The project also occurred in a context of large power disparities among the societal actors. The social actors affected by a research problem, and hoping for positive change, did not have the power to trigger change. Enacting power can enable or hinder transformative actions, hence the agency of the actors involved in TDR is important (House et al. 2024). Such an agency can be individual, in the sense of powerful stakeholders, but could also be grouped in terms of social-self-organisation (Ifejika Speranza et al. 2014; Mukhovi et al. 2020).

Research actors (including researchers) and the research can affect the research context and can be affected by the research context. The examined cases show that the researcher is just one of the actors in a research process. The researcher has own capacities, beliefs, values, attitude, behaviour and agency but is embedded in a social-ecological context (Bergmann et al. 2021). Such contexts can be characterised by interactions with societal actors, other researchers (the research community), own academic organisation, scientific societies and academies, funding organisations and national policies regarding research permits, export control and customs regulations for research equipment. In some contexts, a researcher may have the role of an outsider to the social-ecological reality of everyday living (Lewin 1946), hence requiring adequate time to become accustomed to a research context.

These reflections show that TDR is a social process whose outcomes also depends on actors' concepts of development and well-being and their expectations of how research contributes to them (Schneider et al. 2022). Such expectations affected which research results were valued and put into use by societal actors. In the FoodSAF related assessments, farmers were interested in improving their positions in the value chains, their food security and their incomes through post-harvest management of crops and milk processing. The research participants saw value in the research results that pointed to the need for societal action on improving grain harvest storage and processing milk and benefited from these results by self-organising with the support of the project to make that change happen. These reflections highlight that the degree of desirability of a research outcome depends on the extent to which societal actors are attached to or dependent on them.

Knowledge generated in TDR is expected to be actionable to solve real-world problems. In some instances, actors engaged in the research projects felt gratified that foreigners/outsiders were interested in their knowledge and were willing to share their knowledge. Equally, local actors expected that such research would somehow lead to solving, in their perceptions, more urgent and concrete problem of poverty or immediate food insecurity. In both cases, some societal actors voiced their expectations but also disappointments about knowledge generation and research fatigue. This was through statements regarding other previous research and the mainly socio-economic changes they wish to have, which have either not occurred or to a lower extent than the societal actors expected. This raises questions about whether new knowledge always leads to the expected improvements or societal change. Some knowledge might also be less desirable, while other knowledge remains shielded from society (Ifejika Speranza et al. 2010; Schneider et al. 2022). Several reasons have been reported to explain research fatigue and how it can be addressed (Ashley 2021; Patel et al. 2020). Among them, a TDR approach offered a way to reduce research fatigue: societal actors were involved throughout the research process, and through results exchange workshops and transformative pilot actions that address the immediate societal problems, the benefits of the research could be demonstrated.

TDR also raises questions about a researcher's own expectations and societal expectations on researchers, and about researcher roles and identities—as a scientist, as an activist, as a science-policy-practice dialogue practitioner, scientific adviser or even as a policy maker. Hence, the need to balance engagement in science-societal dialogue, while contributing scientific evidence remains. While it is challenging for researchers to achieve lasting social-ecological change during a project, identifying and communicating solution pathways may be a way to define a researcher's role.

In terms of exploring and validating research impact, the TPAs went beyond identifying and communicating solution options, for instance, by serving as a tool to immediately test/validate scientific results in real-world settings. It supported collective action to address identified challenges through empirical evidence and through a participatory process with stakeholders. However, TPAs might be short-lived, and their effects might not last. Yet, the TPAs in Kenya have potential to bear long-term benefits due to them having been anchored in the County government line ministries and supported by extension personnel (Mukhovi et al. 2024). The TPA idea builds on (participatory) action research approaches and has been previously implemented as Partnership Actions for Mitigating Syndromes (PAMS) to test and validate research findings through short-term partnership actions of researchers and societal actors to address non-sustainable development and to "bridge the gap between research and development" (Messerli et al. 2007: 7). The TPAs thus offer an approach to implement research results, and bridge science-practice gaps in the short-term compared to research-driven policy change, which can take much longer to achieve. Further, the fact that the TPAs met a need identified in a participatory process provides a momentum for sustainability. Many years after their implementation, it would be insightful to re-visit the TPAs to examine if they continued to support food system sustainability, what adaptations stakeholders have made over time and what their current functioning is. Evaluating the impact of the TPA could capture the livelihood outcomes of the project in both sites. Where then to draw the boundary when exploring research impact and moving on to the next research project is not clear cut but even after many years, a research topic can be revisited to explore whether research assumptions in previous research turned out to be and how.

Finally, the limitations of the scope of social-ecological change that can occur through research projects highlights the importance of long-term partnerships between researchers and practitioners (Clark et al. 2016; Jagosh et al. 2015) and the important role of boundary work and organisations, who could take up research results and continuously mobilise societal actors to act on them (Barnes and Van Laerhoven 2015; Clark et al. 2016). Embedding research into larger development activities in the focus geographical region can address this limitation and support social learning (mutual learning process) (Messerli et al. 2007). Such learning enables knowledge dialogue and new knowledge to develop, thereby triggering changes in values, norms and attitudes that enhance the adoption of identified solution options (Messerli et al. 2007).

1.7 Conclusion

This essay set out to reflect on the process of transdisciplinary research drawing on, for illustrative purposes, a research project focused on pathways aimed at achieving sustainable food systems. The reflections show that several major factors are important in the research process. These factors include resources in terms of time and money and prior activities to enable a co-design by researchers and societal actors. Another factor is the choice and depth of stakeholder engagement, which depended on prior partnerships and trust. Methodologically, the research approaches applied ranged between disciplinary, interdisciplinary and transdisciplinary methods. Among others, these approaches encompassed soil and water analysis, participatory workshops, focus group discussions, household surveys, expert interviews, and lifecycle assessments, with varying degrees of involvement of societal actors in the data collection process and the participatory implementation of research results. The TDR approach through continuous interactions between researchers and societal actors triggered early social learning that continued throughout the project. The case studies analyzed show that TDR has its advantages and disadvantages, and hence research approaches in general, including disciplinary, interdisciplinary to transdisciplinary approaches, may be seen as a continuum with an increasing potential in TDR for social learning and social-ecological transformation.

Acknowledgements We express our gratitude to Frank Mintah for providing feedback on an earlier version of this paper and to all academic and non-academic partners with whom we interacted with during the projects.

Funding This research was carried out within the Swiss Programme for Global Issues on Development (r4d programme) funded by the Swiss Agency for Development and Cooperation (SDC) and the Swiss National Science Foundation (SNSF) [Grant number 400540_152033].

References

Adler C, Hirsch Hadorn G, Breu T, Wiesmann U, Pohl C (2018) Conceptualizing the transfer of knowledge across cases in transdisciplinary research. Sustain Sci 13(1):179–190. https://doi.org/10.1007/s11625-017-0444-2

Ashley F (2021) Accounting for research fatigue in research ethics. Bioethics 35(3):270–276. https://doi.org/10.1111/bioe.12829

Barnes C, Van Laerhoven F (2015) Making it last? Analysing the role of NGO interventions in the development of institutions for durable collective action in Indian community forestry. Environ Sci Policy 53:192–205. https://doi.org/10.1016/j.envsci.2014.06.008

Bergmann M, Schäpke N, Marg O, Stelzer F, Lang DJ, Bossert M, Gantert M, Häußler E, Marquardt E, Piontek FM, Potthast T, Rhodius R, Rudolph M, Ruddat M, Seebacher A, Sußmann N (2021) Transdisciplinary sustainability research in real-world labs: success factors and methods for change. Sustain Sci 16(2):541–564. https://doi.org/10.1007/s11625-020-00886-8

Bessa A, Mwangi J, Wachira J (Directors). (2020) Can research projects contribute to community empowerment?—Farmers' feedback. Video. Produced by Sonja Schenkel. Co-Production r4d programme and Patiti Lab. Digital Storyteller synthesis project of the Swiss Programme for Research on Global Issues for Development (r4d programme). [Video recording]

Clark WC, Tomich TP, Van Noordwijk M, Guston D, Catacutan D, Dickson NM, McNie E (2016) Boundary work for sustainable development: natural resource management at the consultative group on international agricultural research (CGIAR). Proc Natl Acad Sci 113(17):4615–4622. https://doi.org/10.1073/pnas.0900231108

Gustafsson KM, Lidskog R (2018) Boundary organizations and environmental governance: performance, institutional design, and conceptual development. Clim Risk Manag 19:1–11. https://doi.org/10.1016/j.crm.2017.11.001

Hoffmann S, Pohl C, Hering JG (2017) Methods and procedures of transdisciplinary knowledge integration: empirical insights from four thematic synthesis processes. Ecol Soc 22(1):art27. https://doi.org/10.5751/ES-08955-220127

House J, Davis N, Dermody BJ, Van Der Horst H, Praasterink F, Wertheim-Heck S (2024) The politics of transdisciplinary research on societal transitions. Futures 164:103499. https://doi.org/10.1016/j.futures.2024.103499

Ifejika Speranza C, Kiteme B, Ambenje P, Wiesmann U, Makali S (2010) Indigenous knowledge related to climate variability and change: insights from droughts in semi-arid areas of former Makueni District Kenya. Clim Change 100(2):295–315. https://doi.org/10.1007/s10584-009-9713-0

Ifejika Speranza C, Wiesmann U, Rist S (2014) An indicator framework for assessing livelihood resilience in the context of social–ecological dynamics. Glob Environ Chang 28:109–119. https://doi.org/10.1016/j.gloenvcha.2014.06.005

Ifejika Speranza C, Kiteme B (2020) Tool for improving food systems. https://boris.unibe.ch/142845/

Ifejika Speranza C, Kiteme B, Wiesmann UM (2008) Droughts and famines: the underlying factors and the causal links among agro-pastoral households in semi-arid Makueni district, Kenya. Global Environ Change 18(1):220–233. https://boris.unibe.ch/34358/

IPBES (2024) Summary for policymakers of the thematic assessment report on the underlying causes of biodiversity loss and the determinants of transformative change and options for achieving the 2050 vision for biodiversity of the intergovernmental science-policy platform on biodiversity and ecosystem services. In: O'Brien K, Garibaldi L, Agrawal A, Bennett E, Biggs R, Calderón Contreras R, Carr E, Frantzeskaki N, Gosnell H, Gurung J, Lambertucci, S, Leventon, J, Liao C, Reyes García V, Shannon L, Villasante S, Wickson F, Zinngrebe Y, Perianin L (eds) IPBES secretariat, Bonn, Germany. https://doi.org/10.5281/zenodo.11382230

Jacobi J, Mukhovi S, Llanque A, Giger M, Bessa A, Golay C, Speranza CI, Mwangi V, Augstburger HF, Bürgi E, Haller T, Kiteme BP, Burgoa JMF Tribaldos D, TM, Rist S (2020) A new understanding and evaluation of food sustainability in six different food systems in Kenya and Bolivia. Sci Rep 10(1):19145. https://boris.unibe.ch/149438/

Jagosh J, Bush PL, Salsberg J, Macaulay AC, Greenhalgh T, Wong G, Cargo M, Green LW, Herbert CP, Pluye P (2015) A realist evaluation of community-based participatory research: partnership synergy, trust building and related ripple effects. BMC Public Health 15(1):725. https://doi.org/10.1186/s12889-015-1949-1

Karrasch L, Grothmann T, Michel TA, Wesselow M, Wolter H, Unger A, Wegner A, Giebels D, Siebenhüner B (2022) Integrating knowledge within and between knowledge types in transdisciplinary sustainability research: seven case studies and an indicator framework. Environ Sci Policy 131:14–25. https://doi.org/10.1016/j.envsci.2022.01.014

Kiatkoski Kim M, Douglas MM, Pannell D, Setterfield SA, Hill R, Laborde S, Perrott L, Álvarez-Romero JG, Beesley L, Canham C, Brecknell A (2022) When to use transdisciplinary approaches for environmental research. Front Environ Sci 10:840569. https://doi.org/10.3389/fenvs.2022.840569

Kläy A, Zimmermann AB, Schneider F (2015) Rethinking science for sustainable development: reflexive interaction for a paradigm transformation. Futures 65:72–85. https://doi.org/10.1016/j.futures.2014.10.012

Larson LR, Cooper CB, Futch S, Singh D, Shipley NJ, Dale K, LeBaron GS, Takekawa JY (2020) The diverse motivations of citizen scientists: does conservation emphasis grow as volunteer participation progresses? Biol Cons 242:108428. https://doi.org/10.1016/j.biocon.2020.108428

Lewin K (1946) Action research and minority problems. J Soc Issues 2(4):34–46. https://doi.org/10.1111/j.1540-4560.1946.tb02295.x

Llanque A, Jacobi J, Tribaldos TM, Mukhovi S, Silvestre C, Tecchio A, Fernandez L, Delgado F, Kiteme B, Maluf R, Bessa A, Speranza CI, Rist S (2021) Transformations towards food sustainability using the participatory food sustainability assessment framework (FoodSAF). Soc Innovat J 5. https://boris.unibe.ch/154390/

Lux A, Schäfer M, Bergmann M, Jahn T, Marg O, Nagy E, Ransiek A-C, Theiler L (2019) Societal effects of transdisciplinary sustainability research—how can they be strengthened during the research process? Environ Sci Policy 101:183–191. https://doi.org/10.1016/j.envsci.2019.08.012

Makri A, Patil M (2021) Food academy: testing sustainability in the system for small farms. https://k4d.ch/food-academy-testing-sustainability-in-the-system-for-small-farms/

McGinnis MD, Ostrom E (2014) Social-ecological system framework: initial changes and continuing challenges. Ecol Soc 19(2):art30. https://doi.org/10.5751/ES-06387-190230

Messerli P, Salmi A, Herweg K, Pfister F, Breu T (2007) 2007. Bridging research and development: partnership actions for mitigating syndromes (PAMS) in the Swiss national Centre of competence in research (NCCR) North-South—Capitalising on experience. NCCR North-South Dialogue, 17. Bern, Switzerland: NCCR North-South

Meyer N, Wehrli D (2024) Final evaluation of the r4d programme. https://www.r4d.ch/api/download/Imh0dHBzOi8vc3RvcmFnZS5nb29nbGVhcGlzLmNvbS9zbmYtbWVkaWEvcjRkLXByb2R1Y3Rpb25fWno1Tlk1SHExT1pRYldSZy9zVmZ1Znh4dlFxbzZNckhjIg==

Mukhovi SM, Kiteme B, Mwangi J, Wambugu G (2024) Transdisciplinary knowledge co-production as a catalyst for community-led innovation: a case study of farmers' milk cooperative in Laikipia Kenya. Front Sustainable Food Syst 8:1494692. https://doi.org/10.3389/fsufs.2024.1494692

Mukhovi SM, Jacobi J, Llanque A, Rist S, Delgado F, Kiteme B, Speranza CI (2020). Social self-organization and social-ecological resilience in food systems: lessons from smallholder agriculture in Kenya and indigenous Guaraní Commun Bolivia. Food Stud 10(1):19–42. https://boris.unibe.ch/142824/

Norström AV, Cvitanovic C, Löf MF, West S, Wyborn C, Balvanera P, Bednarek AT, Bennett EM, Biggs R, Brémond AC, Campbell BM, Canadell JG, Carpenter SR, Folke C, Fulton EA, Gaffney O, Gelcich S, Jouffray J-B, Leach M, Le Tissier M, Martín-López B, Louder E, Loutre M-F, Meadow AM, Nagendra H, Payne D, Peterson GD, Reyers B, Scholes R, Ifejika Speranza C, Spierenburg M, Stafford-Smith M, Tengö M. van der Hel S, van Putten I, Österblom H (2020) Principles for knowledge co-production in sustainability research. Nat Sustain 3(3):182–190. https://doi.org/10.1038/s41893-019-0448-2

Oberlack C, Breu T, Giger M, Harari N, Herweg K, Mathez-Stiefel S-L, Messerli P, Moser S, Ott C, Providoli I, Tribaldos T, Zimmermann A, Schneider F (2019) Theories of change in sustainability science: understanding how change happens. GAIA—Ecolo Perspect Sci Soc 28(2):106–111. https://doi.org/10.14512/gaia.28.2.8

Ostrom E (2011) Background on the institutional analysis and development framework. Policy Stud J 39(1):7–27. https://doi.org/10.1111/j.1541-0072.2010.00394.x

Ott C, Kiteme B (2016) Concepts and practices for the democratisation of knowledge generation in research partnerships for sustainable development. Evidence Pol 12(3):405–430. https://doi.org/10.1332/174426416X14700793045951

Pärli R (2023) How input, process, and institutional factors influence the effects of transdisciplinary research projects. Environ Sci Policy 140:80–92. https://doi.org/10.1016/j.envsci.2022.11.019

Patel SS, Webster RK, Greenberg N, Weston D, Brooks SK (2020) Research fatigue in COVID-19 pandemic and post-disaster research: causes, consequences and recommendations. Disaster Prevent Manage: Int J 29(4):445–455. https://doi.org/10.1108/DPM-05-2020-0164

Pohl C, Klein JT, Hoffmann S, Mitchell C, Fam D (2021) Conceptualising transdisciplinary integration as a multidimensional interactive process. Environ Sci Pol 118:18–26. https://doi.org/10.1016/j.envsci.2020.12.005

Pohl C, Krütli P, Stauffacher M (2017) Ten reflective steps for rendering research societally relevant. GAIA—Ecol Persp Sci Soc 26(1):43–51. https://doi.org/10.14512/gaia.26.1.10

Rigolot C (2020) Transdisciplinarity as a discipline and a way of being: complementarities and creative tensions. Human Soc Sci Commun 7(1):100. https://doi.org/10.1057/s41599-020-00598-5

Rist S, Chiddambaranathan M, Escobar C, Wiesmann U (2006) It was hard to come to mutual understanding …"—The multidimensionality of social learning processes concerned with sustainable natural resource use in India, Africa and Latin America. Systemic Practice Action Res 19(3):219–237. https://doi.org/10.1007/s11213-006-9014-8

Rist S, Golay C, Bürgi E, Burgoa FD, Kiteme BP, Haller T, Speranza CI (2016) Towards food sustainability: reshaping the coexistence of different food systems in South America and Africa? Project description (Towards Food Sustainability Working Paper, Vol. 1). Centre for Development and Environment (CDE), University of Bern. https://boris.unibe.ch/92603/

Rist S, Delgado F, Mukhovi S, Giger M, Llanque A, Bürgi E, Bessa A., Ifejika Speranza C, Jacobi J (2021) Manual for participatory food system sustainability assessments and transformation (FoodSAT)? Steps towards food democracy. Centre for development and environment (CDE), University of Bern. https://boris.unibe.ch/159940/

Schneider F, Llanque-Zonta A, Andriamihaja OR, Andriatsitohaina RNN, Tun AM, Boniface K, Jacobi J, Celio E, Diebold CL, Patrick L, Latthachack P, Llopis JC, Lundsgaard-Hansen L, Messerli P, Mukhovi S, Tun NN, Rabemananjara ZH, Ramamonjisoa BS, Thongmanivong S, Vongvisouk T, Thongphanh D, Myint W, Zaehringer JG (2022) How context affects transdisciplinary research: insights from Asia Africa and Latin America. Sustain Sci 17(6):2331–2345. https://doi.org/10.1007/s11625-022-01201-3

Snapp SS, Bezner Kerr R, Bybee-Finley A, Chikowo R, Dakishoni L, Grabowski P, Lupafya E, Mhango W, Morrone VL, Shumba L, Kanyama-Phiri G (2023) Participatory action research generates knowledge for sustainable development goals. Front Ecol Environ 21(7):341–349. https://doi.org/10.1002/fee.2591

Ustundag E (2022) Transdisciplinarity. In: Showing theory to know theory. In: Ballamingie Pm, Szanto D (eds). https://ecampusontario.pressbooks.pub/showingtheory/

Wiesmann UM, Hurni H, Ott C, Zingerli C (2011) Combining the concepts of transdisciplinarity and partnership in research for sustainable development. University of Bern. https://doi.org/10.7892/BORIS.9026

Chinwe Ifejika Speranza Professor of Geography and Sustainable Development with focus on Land Systems and Sustainable Land Management, Institute of Geography, University of Bern, Bern, Switzerland. Her research and teaching aim to advance understanding of the interactions between land use and management, natural resources and ecosystem functioning, and how to foster sustainability in land systems.

Stellah Mikalitsa Mukhovi Associate Professor in Agricultural Geography, Department of Geography, Population and Environmental Studies, University of Nairobi, Nairobi, Kenya. Her research interest and training is in food systems, social-ecological resilience, food security, livelihoods, climate change and agriculture

Boniface Peter Kiteme Director, Centre for Training and Integrated Research in ASAL Development (CETRAD), Nanyuki, Kenya. His main research interests are in food systems sustainability, integrated water resources management and governance and sustainable land management in the rangelands

Chapter 2
Integral Hope

A Transdisciplinary Model for Human and Environmental Flourishing

Andreas M. Krafft

Abstract This chapter introduces a transdisciplinary model that integrates the core properties of hope and connects them to various realms, from the individual to the interpersonal, collective, environmental, and spiritual. Building on previous psychological research on hope, it outlines the rationale for adopting a transdisciplinary approach and examines the fundamental elements shared by nearly all forms of hope from a philosophical standpoint. The chapter further elaborates on the three universal aspects of hope—wishing, believing in its realization, and trusting oneself and others—and presents a model that illustrates how these basic elements manifest across five interconnected layers. In doing so, it explores the ways in which hope can be expanded and nurtured at each new level. The chapter concludes with the main findings and their implications for human and environmental flourishing.

Keywords Hope · Beliefs · Trust · Collective hope · Environmental hope · Spirituality · Transdisciplinarity

2.1 Different Facets of Hope

Hope is an existential, universal, and manifold phenomenon of human experience that helps people thrive and flourish (Slezackova and Krafft 2017). People hope for many different things in diverse situations and draw nourishment for their hopes from numerous sources (Krafft and Walker 2018). Several disciplines, including theology, philosophy, health sciences, and psychology, have explored the multifaceted nature of hope through distinct thought and academic traditions. Theology, predominantly Christian theology, focuses on the relationship between people's hope and the Divine. Philosophy has mainly sought to identify the key elements that characterize hope and

---

A. M. Krafft (✉)
Institute of Systemic Management and Public Governance, University of St. Gallen, St. Gallen, Switzerland
e-mail: andreas.krafft@unisg.ch

T. Karcher SJ et al. (eds.), *Making Hope Possible*, Environmental Humanities: Transformation, Governance, Ethics, Law 7,
https://doi.org/10.1007/978-3-032-17719-3_2

distinguish it as a human virtue (for an overview, see Blöser 2020). Health sciences, particularly nursing research, examine the hopes of people facing severe illnesses (Herth 2005; Olsman 2020). In positive psychology, hope has been conceptualized as an individual's character strength and a positive emotion that aids people in living a fulfilling life, primarily helping them overcome dire situations (Fredrickson 2004; Peterson and Seligman 2004).

Throughout the centuries, various conceptualizations and definitions have highlighted different aspects, elements, and sources of hope, underscoring its complexity and multidimensionality (Blöser 2020). Receptivity theories suggest that hope is a transcendent phenomenon infused by a higher power, alongside faith and love. On the other hand, agency theories have centered their conceptualization of hope on an individual's capacity to define goals, willpower, and the various pathways a person relies on to achieve those goals (Snyder 2002). Besides the cognitive-individual and the religious-transcendent perspectives, contemporary philosophers and psychologists emphasize the distinctively emotional, social, and collective character of hope, a view supported by the latest advancements in neuroscience (Callina et al. 2018).

According to Fredrickson (2004), hope's positive emotional effect broadens the mindset, nurturing psychological, social, and physical resources to cope with adversity. Hope can be a virtue because it enables people to flourish, allowing them to find meaning in life, foster personal growth, and cultivate positive relationships with others (Cohn and Fredrickson 2009). Because of this broadening and growth effect, hopeful people tend to display more altruistic and generative behavior by taking a long-term view of things, rather than satisfying short-term needs, thinking beyond the struggles of the present moment, and adopting moral values to help others such as friendship, gratitude, generativity, selflessness, kindness, and inclusiveness towards strangers (Cohn and Fredrickson 2006).

In their handbook on virtues and character strengths, Peterson and Seligman (2004) included hope as belonging to the virtue of transcendence. Hope belongs to the virtue of transcendence because it goes beyond one's knowledge and coping capabilities. It enables individuals to build connections to something bigger than themselves, providing meaning, purpose, and fundamental beliefs. Hope is linked to other character strengths such as gratitude, appreciation of beauty and excellence, and spirituality, and is related to values that offer a moral framework, keeping the person committed to the expectation and pursuit of goodness.

A central question is how the objects, targets, and states of affairs that people hope for may influence the way they hope, the sources of hope they draw on, their actions, and the thoughts and feelings associated with hope (Krafft et al. 2023a). People can hold various types of hopes, including material goods, personal achievements, hedonic pursuits, interpersonal relationships, and altruistic motives, which may also influence their overall perception of hope (Averill and Sundararajan 2005). It may not be the same to hope for achieving a specific career goal, for a happy family, or transcendent religious and spiritual experiences. These insights suggest that hoping for certain things can affect how people in different circumstances and environments approach hope, and, therefore, the quality of hope itself (Averill et al. 1990).

The diverse conceptions of hope developed by scientists from diverse disciplines can be complementary and enriching. All these considerations underscore the necessity of integrating existing concepts and elements of hope from different research fields, each highlighting a specific aspect or feature of hope in its own right. Due to its complexity, hope research must be inter- and transdisciplinary, merging theories and research findings from psychology, philosophy, theology, and other disciplines into new concepts and models. As Colla et al. (2022) expound, hope is an experience or phenomenon that emerges from the interplay of individual and contextual factors, which can be modeled as a nested ecological system with different layers. This means we must examine the multi-directional connections between these various layers, such as the individual, the social, the environmental, and the spiritual.

2.2 Basic Elements of Hope

Contemporary philosophers have identified a first common foundation for conceptualizing hope. The so-called standard account or orthodox definition of hope asserts that it involves at least two basic elements: (1) A wish or desire for a valued good or outcome (we only hope for what is important to us); and (2) the belief that the realization of what we desire is possible but not certain (Day 1969; Meirav 2009). Furthermore, contemporary philosophers share a common understanding that these two fundamental elements of the standard definition—desire and possibility—can be regarded as necessary but not sufficient conditions for explaining the phenomenon of hope. Several authors have proposed different features that account for this missing factor. We adhere to numerous psychological and philosophical explanations that define an attitude of substantial trust as a core element of hope (Braithwaite 2004; Erikson 1998; McGeer 2008; Tennen et al. 2002).

Therefore, we propose to understand hope in its fundamental sense as the human capacity that emerges from the combination of *a wish or desire* for a valued outcome or state of affairs, along with the *belief* that its realization is possible—albeit uncertain and not necessarily likely—and the *trust* in the availability of some internal or external resources, either existing or future, that could help facilitate its realization, especially when confronting obstacles and setbacks (Krafft et al. 2023a). The willpower to act and persist stems from the nature and importance of the desired good, as well as the intensity of the belief and trust held by the person who hopes for it.

These fundamental elements should capture the essence of hope while being broad enough to be applied in various situations, across different levels (e.g., individual, interpersonal, social, and transcendent), and within diverse cultural contexts. An additional advantage of this conceptualization of hope is that it can be applied to two distinct forms of hope, which in the literature are known as the form of specific hopes and the form of general or fundamental hopefulness (Calhoun 2018; Dufault and Martocchio 1985; Godfrey 1987; Shade 2001): (1) Specific hopes are directed toward concrete outcomes (goals, events, circumstances) connected with the belief in the possibility of their fulfillment and linked to particular resources for their realization; (2) Fundamental hope is usually oriented toward a good life in general. It might be based on the overall belief in the world's goodness and relies on a basic trust in one's capabilities or some external power, fortune, or fate.

Drawing on existing publications from diverse scientific backgrounds, the following sections provide an overview of the three fundamental elements of hope, as conceptualized in this chapter.

2.3 Wish, Belief, and Trust

Wishes and desires, the first element inciting hope, evolve from what people value and want to happen because it is significant to them. To hope for something always implies a conscious or unconscious motive, the reason for hope, which reflects the values people are committed to (Blöser and Stahl 2017). We hope for something when what we hope for has meaning and value. Since hope is a strong motivator for action, the first task must be to pay attention to the quality, meaning, and significance of the targets of hope. People in different circumstances may hope for diverse things, including good health, a fulfilling job, a happy family, fame, and a fulfilling life, and rank their hopes according to their relative value, creating a hierarchy of hopes (Krafft et al. 2023b; Shade 2001; Shin et al. 2013). However, hope can, but need not, involve a concrete goal or personal achievement, for example, when people hope for peace or the well-being of others. Therefore, the formulation of hoped-for ends and the conscious assessment of the desirability of specific hopes are of utmost importance to understand the lived experience of hope (Shade 2001).

A central question is how the objects, targets, and states of affairs people hope for may influence the sources of hope they draw on, their actions, and the thoughts and feelings associated with hope (Olsman 2020). Different targets of hope can be linked to culturally transmitted worldviews, values, and norms. Culture contains implicit or explicit assumptions about what is good and right. The prevailing values in a particular culture, such as tolerance, care, creativity, power, or performance, can influence people's hopes. For example, Singelis et al. (1995) distinguished individualist and collectivist cultures based on people's goals. People in individualistic societies tend to hold and pursue more self-centered goals that reflect personal wishes. In more collectivist cultures, individuals tend to prioritize goals that align with the desires and wishes of their family and immediate social environment, including maintaining good social relationships, harmony with their surroundings, and supporting others (Triandis 1997). In one case, the focus is on fulfilling one's wishes and desires, achieving personal success, and experiencing novel adventures, with goals requiring personal effort to be accomplished. In other cases, individuals prioritize tolerance, respect, and care for others over personal wishes, goals, and individual performance (Eid and Diener 2009).

The second element in the basic definition of hope is the *belief* that achieving the desired good is possible, even in the face of uncertainty and difficulty. Since the future does not exist and we cannot know it, it remains a realm of speculation and belief (Slaughter 1993). To hope is to stay open to the future and to believe in how things could be rather than how they currently are. In classical cognitive psychological theories, it is assumed that when we believe in ourselves, based on experience, we also think that what we desire is more likely to occur (Snyder 1994; Stotland 1969). Furthermore, hope is grounded in the belief in one's capabilities to overcome obstacles and setbacks. From more recent psychological and philosophical perspectives, neither anticipating a high probability of fulfillment nor having personal control over the hoped-for outcome is a prerequisite for hope (Bruininks and Malle 2005; Milona 2020). Hope is distinct from positive future expectations in that it pertains to believing in even a slight possibility of achieving a particular wish (Krafft et al. 2021; Scioli et al. 1997). Moreover, the belief in the possibility of a specific event is not founded on evidence and objective facts; rather, it is predominantly of a subjective nature (Stockdale et al. 2019). Holding a belief that enables an individual to hope despite negative evidence is neither a sign of self-deception nor an illusion regarding how things appear. Moreover, hope is rooted in the belief that, especially in times of trial, the future will present new possibilities that we cannot yet imagine or contemplate (Lear 2006).

According to Meirav (2009), to understand the true nature of hope, we must recognize that hope is always related to some external factor beyond our resources or agency. The external factor account comprises the belief that the hoped-for good is possible but uncertain, arduous, and not entirely within one's control (Miceli and Castelfranchi 2010). This external factor can be another person or group of individuals, such as our own family; it could be related to social institutions, like the government or employers, but it can also be the belief in fate, luck, or a transcendent higher power. What is central is that, to be hope-sustaining, the external factor must be conceived as favorable to one's values, interests, and wishes (Meirav 2009). Hope is rational and appropriate as long as one believes in the goodness of the external factor. Pragmatist hope is grounded in the conviction that believing in life's benevolence is justified and that, through collective efforts, we can continually make the world a better place (Blöser and Stahl 2017). It is essential to note that relying on external forces is not necessarily a sign of passivity, disengagement, or wishful thinking. Still, it can be experienced as an expansion and empowerment of one's agency, commitment, and willingness to act (Shade 2001).

In a broader sense, beliefs are constitutive elements of worldviews, i.e., assumptions about the nature, quality, and meaningfulness of the world, as well as why it is as it is and how it should be. These fundamental beliefs are theories or cultural narratives that guide our thoughts, emotions, and behaviors, particularly in anticipating future events (Janoff-Bulman 1989, 1992). These worldviews are especially significant when a person faces a stressful situation or experiences despair. For example, basic world beliefs pertain to the quality and fundamental character of human nature, determining whether it is viewed as good or evil, and whether the world is perceived as just or unjust (Clifton et al. 2019; Lerner 1980). Hope is therefore heavily influenced by cultural beliefs, assumptions, and attitudes that shape our perceptions about the world and ourselves, as well as our behaviors (Ibrahim 1984; Janoff-Bulman 1992; Koltko-Rivera 2004).

Beliefs and desires can influence one another. The meaning, value, and importance of a particular desire shape the belief that this desire is at least possible. Furthermore, if one believes that the fulfillment of a vital wish is possible, one desires it even more fervently (Milona 2019). In contrast, believing in the impossibility or difficulty of a wish or desire may lead the individual to adopt a passive attitude toward it, or even to renounce it altogether. This means that the affective state of hopelessness can be transformed, on the one hand, by emphasizing the significance of the desired outcome, and on the other, by modifying the individual's beliefs, evaluations, and judgments. Sometimes, it's merely a matter of reframing and viewing the problem from a different perspective (Ojala 2012).

Several authors agree that hope does not coincide with positive expectations about a probable outcome and that uncertainty is a constitutive element of hope, particularly about realizing the hoped-for end and one's self-efficacy (Miceli and Castelfranchi 2010). But then, what gives people the power to believe, for example, in healing, despite an unfavorable prognosis? What strengthens the belief in a positive outcome despite all negative evidence? Some authors emphasize individuals' capabilities (Snyder 1994), while others highlight the importance of social support from family members or close friends (Scioli and Biller 2009). Still, others highlight faith in a benevolent higher power (Scioli 2007) to whom one can revert when facing critical life situations.

The third element in our conceptualization of hope focuses on the various forms and expressions of *trust* related to resources that encourage people to believe in the realization of hoped-for outcomes in uncertain circumstances. Trust is the ability to transcend current conditions and to courageously be open to new and unknown solutions. In hoping, we experience the world as open to its possibilities and development and trust the power of our own and others' actions (Webb 2007).

Following the work of Earle and Siegrist (2006), we can distinguish between the concepts of trust and confidence. Confidence is defined as a reason-based assessment of a high probability of achieving a goal, fitting the definition of optimism and positive expectations; trust, on the other hand, is a relational phenomenon supported by social attachment and characterized by shared values such as benevolence, integrity, fairness, and caring. Hope is related to trust rather than confidence, as confidence often renders hope unnecessary (Miceli and Castelfranchi 2010). Trust is closely tied to vulnerability, uncertainty, and faith, thereby fostering interdependence with others.

Trust lies at the core of most sources of hope: trust in our current and future abilities, trust in family members and friends who are ready to support us emotionally and instrumentally, trust in the broader community and social and political institutions, and trust in a Divine Power that is always present and on our side. Individuals in different countries or from diverse social groups may differ in their perceptions of sources of hope (Averill et al. 1990; Averill and Sundararajan 2005). In individualistic societies, hope is grounded in autonomy, independence, and self-efficacy, often relying primarily on oneself, one's capabilities, and commitment. However, at times, there is little one can do to contribute to the realization of one's hopes. People in collectivistic cultures may rely more on the emotional and instrumental support of their family members and closest friends. Religious individuals might place their trust and hope in a benevolent Higher Power.

2.4 A Nested Model of Hope

Beyond the basic definition and elements, hope's complexity has led several researchers to conceptualize it as a multidimensional phenomenon, resulting in a variety of models, including dimensions such as personal mastery, social relationships, collective endeavors, spiritual and religious practices, and several others (Dufault and Martocchio 1985; Farran et al. 1995; Scioli et al. 2011). Research findings from various authors suggest that there is not just one form of hope, but rather many different expressions of hope, varying according to the diverse range of personal, social, and collective wishes, beliefs, and sources of trust. Inspired by systems theory as proposed by Colla et al. (2022) and guided by the idea that different areas and levels of hope influence one another as expressed by Lomas et al. (2015), we propose a model that integrates the three basic elements of hope—wishes, beliefs, and trust—with several interrelated layers represented by a nested model, resulting in different types of hope, ranging from the individual to the transcendent, as presented in Fig. 2.1.

In the following sections, we will describe every layer of hope, present the type of hope distinctive to each layer, and explain the specific qualities of the basic elements

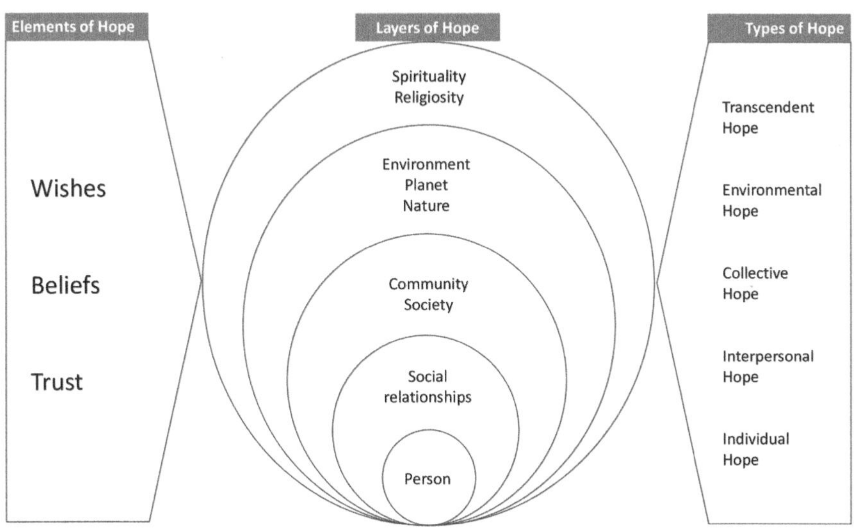

Fig. 2.1 Dimensions of Hope

of hope—wishes, beliefs, and trust—at each level. Furthermore, we will elaborate on how the different layers are interrelated, how they interact with one another, and how each new upper layer contributes to broadening and enhancing the individual, social, and collective capacity to hope.

2.4.1 Person: Individual Hope

In traditional psychological theories, hope is an individual disposition that enables a person to set goals, possess the motivation to commit to achieving those goals, and utilize their abilities to develop pathways to reach them (Snyder 1994; Stotland 1969). These theories are based on self-perception regarding the individual's capabilities to attain specific ends and conceptualize hope as an individual trait or disposition. In this respect, classical psychological hope theories closely relate to concepts such as self-efficacy, self-confidence, and self-esteem (Aspinwall and Leaf 2002).

Dispositional hope involves aspects that the individual can control or at least believes he can. The act of hoping is a mental process that connects one's abilities and actions to the expected outcome. According to Snyder (2000), people who are hopeful feel in control of their lives and take responsibility for their circumstances by actively seeking solutions to their problems. Individual hope encompasses the perception and belief in one's abilities to discover potential ways to achieve goals, especially when obstacles and challenges arise along the path to those goals. Thus, this belief can be linked to coping competence, which is grounded in mental flexibility and creativity. Hopeful individuals tend to be ambitious, setting increasingly demanding goals compared to those who are less hopeful. Self-confidence encompasses beliefs in one's ability to succeed, which is related to persistence, courage, and achievements (White 2009).

As Snyder (1994) acknowledges, dispositional hope is an egocentric phenomenon because it centers on the perception and evaluation of oneself and one's personal goals. Moreover, the dispositional hope theory remains entirely neutral regarding the goodness or moral values of the goals. It does not matter whether a goal aims at something positive or negative or whether it is ethical or legal (Snyder 2002). Goals need only be meaningful and desirable for the person hoping, so they will persist in focusing on them, even when faced with resistance.

The risks of an overly individualistic perspective are that hope can evolve into blind, compulsive, and selfish actions that consume and constrain the individual. Through willful hope, as McGeer (2004) calls it, people can become so focused on achieving their goals that they ignore others and treat them as mere instruments. When the realization of one's goals takes precedence, it can justify all means, regardless of the undesirable consequences that may arise. Individuals driven in this way show little interest in or understanding of the wishes and hopes of others. In a negative sense, willful hope can lead to overestimating oneself and isolating oneself from others.

Put in a broader perspective, we must acknowledge that we cannot always be in control and that we must accept our limitations and dependence on others (Bruininks and Malle 2005). People can still hope, even if they cannot achieve the desired result solely through their efforts. These considerations pave the way for a broader, individual as well as interpersonal, social, cultural, environmental, and spiritual conceptualization of hope and agency.

2.4.2 Social Relationships: Interpersonal Hope

To achieve one's goals, people often need to work with, rely on, and trust others. For Marcel (1951), hope is that quality through which we can see beyond ourselves and our limitations, directing our attention to what is possible with the help of others. At the core of the interpersonal account of hope lies a relational understanding of the self and agency. Hope is not merely a private mental state; the activity of hoping thrives in the interaction between a person and his social environment. Hope is an interpersonal and self-transcendent phenomenon. Other people strengthen our hopes through their encouragement, presence, and active support. We expand our capacity to act by placing our hope in others, whether they are family members, friends, colleagues, or healthcare professionals (Martin 2019). Therefore, hope encompasses not only personal mastery and individual goals but also interpersonal trust and connection with others (McGeer 2004).

The value of social relationships is most evident in critical or deeply personal life situations, such as illness and adversity, where one feels lost and relies on the assistance of others (Marcel 1951). The recognition and acceptance of one's vulnerability, coupled with the conviction that human flourishing is only possible within a community of people who care for one another, make compassion and solidarity a vital element of hope (Kadlac 2015). Social hope begins by accepting one's limitations and those of others, but without surrendering to them. When we trust others, we acknowledge our limitations with humility and recognize the need for support and care. Moreover, it empowers individuals to develop their agency and willpower, as well as their capabilities and possibilities, transcending current limitations while caring for and supporting each other in a process of mutual growth (McGeer 2004, 2008).

Interpersonal hope broadens awareness of what life is truly about and what is worth hoping for (Braithwaite 2004; Stitzlein 2019). The focus is not only placed on an object or a personal goal, but, above all, on another person for whom we hope (Howell and Larsen 2015). By doing so, we convey to them the sense that their wishes and hopes are important to us. Simply by caring about and taking an interest in the hopes of others, we signal to them how valuable their hopes are, which gives their hopes new energy and vitality. This makes people realize how powerful and invigorating mutual attention, encouragement, and support can be for maintaining one's hopes and those of others.

In tough times, people need the availability and supportive affection of others who believe in and for them, especially when they can no longer do so on their own (Howell and Larsen 2015). By believing in them, family and friends reinforce their belief in the realizability of their wishes and encourage them to persevere and not give up (McGeer 2008). When people are there for others, believe in them, encourage them, and stand up for them, this bolsters their hope, faith, and willpower (McGeer 2004). The appreciation that others express through their faith and trust revitalizes their overall capacity for hope and confidence in their abilities. Thanks to the encouragement from others, people can feel motivated again, recognize their strengths with more confidence, and cultivate a positive and hopeful outlook toward the future.

All of this highlights the importance of mutual care and helpfulness in preserving hope. The logical conclusion from these phenomena is that our ability and willingness to hope depend not only on how well we can hope for ourselves but also on how well we can hope with and for others (McGeer 2004; Snow 2019). Hope is a gift that empowers others and, in turn, ourselves. When we respond to the hopes of others and receive their encouragement for our hopes, we participate in their lives and expand our horizons of hope.

2.4.3 Community and Society: Collective Hope

Collective hope represents a distinct form of social hope, characterized by shared visions that aim to positively shape the world toward a meaningful and dignified life within a social community (Braithwaite 2004; Kleist and Jansen 2016). We speak of collective hope when a community of people shares common beliefs and desires, bringing them to fulfillment in a hopeful manner (Blöser and Stahl 2019). Rather than focusing solely on individual objectives, people engage in a larger community to pursue overarching social, economic, and environmental goals (Stahl 2019). From a cultural perspective, individual and social hopes emerge through shared beliefs, which, at a specific time and place in history, constitute collective worldviews (Naderi Farsani and Abolghasemi 2008).

According to pragmatist philosophers such as Dewey (1916/1980) and Rorty (2002), belief in the good in life serves as a prerequisite for a serious commitment to a better world. While a society dominated by fear is characterized by selfishness and mutual competition, a community of hope fosters greater cohesion, solidarity, and cooperation among its members (Stahl 2019). In collective hope, we acknowledge the difficulties of our current circumstances while believing that improvement is possible through our conjoint engagement and efforts. The belief in future possibilities characterizes collective hope and trust in the collective power of determination and action (Rorty 1998). Rorty claims that hope for a better world is the belief in the possibility that "unnecessary human suffering can be decreased, and human happiness thereby increased" (Rorty 2002, p. 154). The hope for a world in which there can be more justice and less suffering is directed towards concrete concerns such as everyone having enough to eat, a dignified home, access to medical care, and education (Rorty 2002).

Pragmatist philosophers emphasize our dependence on others and the necessity of trusting that others will also strive for a common cause, particularly in hopes of peace, justice, or a sustainable society (Stitzlein 2019). What holds societies together are shared hopes of a better collective future (Rorty 2002). Instead of focusing solely on one's individual goals, community members are committed to achieving shared objectives and work together to pursue them (Rorty 1999). This belief enables the emergence of transformative hope, which, according to Webb (2007), involves developing a vision of the future based on the belief in the socially instilled human capacity to improve the world by changing the status quo. In this respect, political hope needs public trust in social and democratic institutions (Huber 2021; Moellendorf 2006).

A community involves not only a common cause but also the respect and empowerment of each member. Cultural beliefs about how things are and how they could and should be are constituents of the individual's perception and volition (Miller 1999). The quality and role of these fundamental cultural beliefs can significantly impact the quality and the sources of individual hope. Belonging to a social community enhances one's life meaning, effectiveness, and self-esteem (McGeer 2004). What we would not be able to achieve alone can be realized through the collective efforts of a community, thanks to the individual contributions of all its members (Braithwaite 2004). With the community's strengthening power, we can expand our capacity for action and hope. A community thrives when its members can make individual contributions, grow and develop their identities, and feel valued and supported by others (Shade 2019). A collective vision and action for the future, with the aspiration for an improved life, connects individual people with others, fosters the common good, and instills a sense of purpose in people's lives (Stitzlein 2019).

Social and voluntary institutions, including aid and non-governmental organizations, as well as schools and churches, can be vibrant communities of hope where individual efforts and actions intersect with those of others. Serving others is a clear manifestation of hope and faith, fostering belief in and encouraging dedication to a better future. On one hand, such institutions inspire hope in societal goals or ideals, and on the other, they empower action. In institutions of hope, people can learn to believe in the future and actively engage in extraordinary endeavors (Braithwaite 2004).

2.4.4 Planet and Nature: Environmental Hope

The ecological crisis, characterized by persistent global warming, pollution of landscapes and waters with toxic waste, environmental catastrophes, and a continuing decline in biodiversity, poses a significant threat to the future of our world. Since these urgent ecological challenges cannot be addressed by individual efforts alone, people are responding to the threats in three distinct ways: (1) denial and indifference, (2) pessimism and despair, and (3) shared concern and hope (Ojala 2015). The first two options, denial and despair, imply a significant degree of helplessness and lack of hope that leads to a vicious cycle and a self-fulfilling prophecy (Lueck 2007): The more people believe in the indifference and shortage of engagement of others, the more they will assume that their actions are in vain and refrain from engaging themselves in pro-environmental actions, and the more they will fall into despondency and pessimism (Kretz 2013). Under these circumstances, environmental scientists are now emphasizing the existential role and importance of hope as a motivational force toward pro-environmental behavior. Environmental hope entails a deep concern for the environment, coupled with the belief that not everything is lost and that a positive change toward a sustainable society and environmental regeneration is still possible (Maartensson and Loi 2022).

Ecological hope begins with recognizing that nature and non-human goods possess intrinsic value, independent of our own needs (Kretz 2013). It comprises a self-transcendence attitude, where the desire for the thriving of the entire planet and caring for all life forms pairs with—and sometimes surpasses—people's self-centered wishes (Nolt 2010). The desire to protect, care for, and treat nature as a higher common good is a prerequisite for overcoming indifference and engaging in meaningful pro-environmental actions. However, the wish alone is not sufficient, as strong concerns and feelings of impotence can drive people into despondency. To overcome hopelessness and transform environmental concern into practical action, it is necessary to reorient people's beliefs in several ways (Ojala 2012). Firstly, it is essential to believe that, despite dire prognoses, it is still possible to mitigate, prevent, and even reverse environmental damage.

To foster the belief in a sustainable future, it is furthermore necessary to reconceptualize the relationship between human beings and nature. A significant point is the belief in the regenerative capacity of nature itself. This implies expanding the notion of human agency by recognizing that nonhuman beings, including animals, plants, and entire ecosystems, possess a specific agency themselves (Northcott 2020). Moreover, it is necessary to believe that one's actions are not futile but will have a positive impact and that we are not doing this alone but that many other people are also committed to environmentally responsible behavior (Lueck 2007). Lastly, we must believe that our well-being and flourishing are significantly enhanced by our connection and attachment to nature, as well as by taking responsibility toward the environment (Chawla 2020; Kerret et al. 2016).

To strengthen this sense of belief in nature, in others, and oneself, it is necessary to develop trust in one's capabilities as well as trust in other societal actors and instances, such as the government, science, environmental organizations, technological industries, and fellow citizens (Ojala 2015). However, this trust will not emerge easily, but only through concrete measures of collective endeavors that result from shared values and visions of the future.

In conclusion, environmental hope serves as the foundation for promoting a healthier environment through individual and collective values, goals, beliefs, trust, and actions, fostering a sense of cohesion and community within society, and enhancing both physical and psychological well-being in individuals. By doing so, environmental hope will have a positive impact on collective, interpersonal, and individual hope.

2.4.5 *Spirituality and Religiosity: Transcendent Hope*

The realm of religious and spiritual hope is ancient and vast, featuring several peculiarities among different religious denominations and creeds. The theological virtue of hope is rooted in faith and trust in a benevolent and almighty God, grounded in His love for humanity and all of His creation (Pinsent 2020). For moral theologians such as Thomas Aquinas, religious hope differs from human hope because it is a gift from God, who is the ultimate source of everything. The opposite of spiritual hope is not only despair, which is a lack of faith in God, but also the presumption of trusting only oneself rather than relating to and trusting God. Religious hope is transcendent and absolute because it is rooted in what cannot be known or seen, and therefore, can only be believed in (Jeffrey 2019). Faith is the assurance of things hoped for because everything good comes from God's power and mercy (Pinsent 2020).

In most religious traditions, transcendent hope is not directed towards material things and selfish goals. In the Buddhist and Hindu traditions, hope is grounded in gratitude, compassion, and detachment from unnecessary earthly desires (Dunlap 2019; Krafft and Choubisa 2018; McDonald 2008). Spiritual hope is directed towards things that transcend the material world, such as liberation from evil, overcoming suffering and death, eternal life, and unity with God. Hope based on God does not refer to what already exists, but is guided by the idea of a better, yet partly still unknown, world, and serves as the trigger for the hoped-for radical transformation of the world (Moltmann 2021).

In monotheistic religions, transcendent hope is determined by the two most salient qualities of an agentic God: omnipotence and love. Through the faith in an almighty God, believers gain the certainty that things that they hope for but cannot see or understand are nevertheless fulfilled (Pinsent 2020). From a spiritual and religious perspective, there is nothing impossible, creating room for an infinite number of new opportunities. The power of faith and hope is rooted in a belief in a better future and trust in a spiritual, creative force that is mysteriously available to people, enabling them to feel supported and empowered. Religious and spiritual hope can become a new life force, especially where people have nothing more to expect.

Furthermore, spiritual hope gains its fundamental character only through faith in God's unconditional love. In most religions, the sacred is characterized by love and compassion. To the contrary, as long as one believes in capricious and punishing Gods, hope will have a fragile basis. Hope, rooted in the love of God, empowers people to persevere in hope even when they feel helpless. In this sense, hope conveys the conviction that, despite all suffering and failure, we are loved and that our lives have meaning and significance (van Vliet 2020). This gives people the courage to stand up again and again, and to persevere.

The belief that hope comes from the Almighty does not mean that religious hope is a passive hope. Hope and faith are not only directed towards the future and the afterlife, but also shape the way people live in this world. As human beings, we are called to receive love and to share it with others (Marcel 1951). Hope anchored in love and faith is combined with positive feelings, expanded awareness, and, above all, the values of mutual understanding, respect, tolerance, appreciation, and helpfulness toward all people (Jeffrey 2019; Osmani 2008). Transcendent hope gives the courage and, above all, the strength to affirm the good, that is, to profess love and to avoid evil, that is, egoism, selfishness, and hardheartedness. Our task is to give hope to others by preventing or reducing their suffering. This motivates religious people to make a personal commitment to the poor, the oppressed, and those in need. For Moltmann (2021), humanity has the mission to work towards the promised world by transforming our own lives and the existing circumstances on earth for the better. Faith in God conveys the belief that the individual is entitled, empowered, and assisted by God to do whatever is necessary to achieve the hoped-for good, thereby expanding and strengthening their individual and collective agency. This faith fosters hope in humanity's ongoing moral development and flourishing.

2.5 Conclusion: Integral Hope for Human and Environmental Flourishing

The current chapter begins by highlighting the multifaceted forms of hope and the various research traditions that have investigated these diverse forms and their general connection to human flourishing. From a positive psychological perspective, we recognize that hope has the property of broadening and building people's orientation toward the future, and that it can be considered a character strength belonging to the virtue of transcendence (Fredrickson 2004; Peterson and Seligman 2004). While philosophers have sought the fundamental properties that define the phenomenon of hope across all its different manifestations, psychologists and health researchers have identified several dimensions that are characteristic of hope in particular situations. The purpose of this chapter was to identify and describe the fundamental properties common to (almost) all diverse expressions of hope; elaborate a transdisciplinary, systemic, and nested model that describes different forms and levels of hope, along with the specific characteristics of the basic elements of hope at each level, and the interaction of the different hope forms with each other; and propose that each level contributes to broaden and build the capacity to, and the effectiveness of, hope for a flourishing life, society, and planet.

Hope fundamentally refers to the human ability that arises from a wish for a desired outcome or situation, coupled with the belief in the possibility of its occurrence—though uncertain and not necessarily probable—and the confidence in the presence of some internal or external resources, whether currently available or anticipated, that could assist in achieving it, particularly when facing challenges and failures. Desires, beliefs, and trust can relate to various levels or layers: personal, interpersonal within immediate social relations, collective at the community or societal level, environmental concerning the planet and nature, and transcendent linked with spirituality and religion. Our empirical research has shown that individuals who find meaning in personal growth, relationships with family and friends, community engagement, sustainability, and religious faith experience significantly higher levels of hope and well-being compared to those who lack one or more of these sources of meaning (Krafft 2024).

This leads us to reconsider the classical psychological understanding of hope and agency. The emotions, character traits, and behaviors of an individual are inseparably connected to their relationships with their closer and broader social, natural, and even supernatural environment. Moreover, it is questionable whether the distinction between the individual and the physical and metaphysical environment, in the classical Western sense, can still be maintained (Heidegger 1953, 2010). Concerning hope and agency, we argue that the individual and the social, natural, and spiritual realms interact, enhancing and expanding one another. Individual hope and agency are fostered and, in turn, nurture interpersonal, collective, environmental, and transcendent hope. Human flourishing can best take place when individuals can develop a sense of self-confidence and hope for the future by feeling attached to and supported by loved family members and friends, by being connected to and engaging with other people in a common cause for their community or society, feeling energized by the power of nature, and engaging in its restoration, and feeling empowered by a benevolent and almighty higher power. Our task is to transmit this message to as many people as possible and to support people and communities in finding hope for themselves and others.

References

Aspinwall LG, Leaf SL (2002) In search of the unique aspects of hope: pinning our hopes on positive emotions, future-oriented thinking, hard times, and other people. Psychol Inq 13(4):276–288. https://www.jstor.org/stable/1448868

Averill JR, Sundararajan L (2005) Hope as rhetoric: cultural narratives of wishing and coping. In: Eliott J (ed) Interdisciplinary perspectives on hope. Nova Science, pp 133–165

Averill JR, Catlin G, Chon KK (1990) Rules of hope: recent research in psychology. Springer. https://doi.org/10.1007/978-1-4613-9674-1_1

Blöser C (2020) Enlightenment views of hope. In: van den Heuvel S (ed) Historical and multidisciplinary perspectives on hope. Springer, pp 61–76. https://doi.org/10.1007/978-3-030-46489-9_4

Blöser C, Stahl T (2017) Hope. In: Zalta E (ed) Stanford encyclopedia of philosophy. Online version. Center for the Study of Language and Information

Blöser C, Stahl T (2019) An introduction. In: Blöser C, Stahl T (eds) The moral psychology of hope. Rowman & Littlefield International, pp 1–11

Braithwaite V (2004) Collective hope. Ann Am Acad Pol Soc Sci 592(1):6–15. https://doi.org/10.1177/0002716203262049

Bruininks P, Malle B (2005) Distinguishing hope from optimism and related affective states. Motiv Emot 29(4):324–352. https://doi.org/10.1007/s11031-006-9010-4

Calhoun C (2018) Doing valuable time: the present, the future, and meaningful living. Oxford University Press

Callina KS, Snow N, Murray ED (2018) The history of philosophical and psychological perspectives on hope: toward defining hope for the science of positive human development. In: Gallagher MW, Lopez SJ (eds) The Oxford handbook of hope. Oxford University Press, pp 9–26. https://doi.org/10.1093/oxfordhb/9780199399314.013.2

Chawla L (2020) Childhood nature connection and constructive hope: a review of research on connecting with nature and coping with environmental loss. People Nat 2(3):619–642. https://doi.org/10.1002/pan3.10128

Clifton JD, Baker JD, Park CL, Yaden DB, Clifton AB, Terni P, Miller JL, Zeng G, Giorgi S, Schwartz HA, Seligman ME (2019) Primal world beliefs. Psychol Assess 31(1):82–99. https://doi.org/10.1037/pas0000639

Cohn MA, Fredrickson BL (2006) Beyond the moment, beyond the self: shared ground between selective investment theory and the broaden-and-build theory of positive emotions. Psychol Inq 17(1):39–44. https://www.jstor.org/stable/20447297

Cohn MA, Fredrickson BL (2009) Positive emotions. Lopez S, Snyder CR (eds) Oxford handbook of positive psychology, vol 2. Oxford University Press, pp 13–24

Colla R, Williams P, Oades LG, Camacho-Morles J (2022) "A new hope" for positive psychology: a dynamic systems reconceptualization of hope theory. Front Psychol 13:809053. https://doi.org/10.3389/fpsyg.2022.809053

Day JP (1969) Hope. Am Philos Q 6(2):89–102. https://www.jstor.org/stable/20009295

Dewey J (1916/1980) Democracy and education. Project Gutenberg

Dufault K, Martocchio BC (1985) Hope: its spheres and dimensions. Nurs Clin North Am 20(2):379–391. https://doi.org/10.1016/S0029-6465(22)00328-0

Dunlap R (2019) A Zen Buddhist conception of hope in enlightenment. In: Blöser C, Stahl T (eds) The moral psychology of hope. Rowman & Littlefield, pp 209–226

Earle TC, Siegrist M (2006) Morality information, performance information, and the distinction between trust and confidence. First J Appl Soc Psychol 36(2):383–416. https://doi.org/10.1111/j.0021-9029.2006.00012.x

Eid M, Diener E (2009) Norms for experiencing emotions in different cultures: inter-and intranational differences. In: Diener E (ed) Culture and well-being. Springer, pp 169–202. https://doi.org/10.1007/978-90-481-2352-0_9

Erikson EH (1998) The life cycle completed (extended version). WW Norton & Company

Farran CJ, Herth KA, Popovich JM (1995) Hope and hopelessness: critical clinical constructs. Sage

Fredrickson BL (2004) The broaden-and-build theory of positive emotions. Philos Trans-R Soc Lond Ser B Biol Sci 359(1449):1367–1378. https://doi.org/10.1098/rstb.2004.1512

Godfrey JJ (1987) A philosophy of human hope. Martinus Nijhoff. https://doi.org/10.1007/978-94-009-3499-3

Heidegger M (1953/2010) Being and time. Suny Press

Herth K (2005) State of the science of hope in nursing practice: hope, the nurse, and the patient. In: Eliott JA (ed) Interdisciplinary perspectives on hope. Nova Science Publishers, pp 169–211

Howell AJ, Larsen DJ (2015) Understanding other-oriented hope: an integral concept within hope studies. Springer International Publishing. https://doi.org/10.1007/978-3-319-15007-9

Huber J (2021) Defying democratic despair: a Kantian account of hope in politics. Eur J Polit Theo 20(4):719–738. https://doi.org/10.1177/1474885119847308

Ibrahim FA (1984) Cross-cultural counseling and psychotherapy: an existential-psychological approach. Int J Adv Couns 7(3):159–169. https://doi.org/10.1007/BF00116188

Janoff-Bulman R (1989) Assumptive worlds and the stress of traumatic events: applications of the schema construct. Soc Cogn 7(2):113–136. https://doi.org/10.1521/soco.1989.7.2.113

Janoff-Bulman R (1992) Shattered assumptions: towards a new psychology of trauma. Am J Clin Hypn 36(3):222–225. https://doi.org/10.1080/00029157.1994.10403078

Jeffrey A (2019) Hope in Christianity. In: Blöser C, Stahl T (eds) The moral psychology of hope. Rowman & Littlefield, pp 37–55. https://doi.org/10.5771/9781786609731

Kadlac A (2015) The virtue of hope. Ethical Theory Moral Pract 18(2):337–354. https://doi.org/10.1007/s10677-014-9521-0
Kerret D, Orkibi H, Ronen T (2016) Testing a model linking environmental hope and self-control with students' positive emotions and environmental behavior. J Environ Educ 47(4):307–317. https://doi.org/10.1080/00958964.2016.1182886
Kleist N, Jansen S (2016) Introduction: hope over time—crisis, immobility, and future-making. Hist Anthropol 27(4):373–392. https://doi.org/10.1080/02757206.2016.1207636
Koltko-Rivera ME (2004) The psychology of worldviews. Rev Gen Psychol 8(1):3–58. https://doi.org/10.1037/1089-2680.8.1.3
Krafft AM (2024) Our hopes, our fears, our future. Unpublished keynote presentation at the 11th European conference on positive psychology ECPP, Innsbruck, Austria
Krafft A, Choubisa R (2018) Hope in the Indian psychology context: philosophical foundations and empirical findings. In: Krafft A, Perrig-Chiello P, Walker A (eds) Hope for a good life. Results of the hope-barometer international research program. Springer, pp 131–163. https://doi.org/10.1007/978-3-319-78470-0_7
Krafft AM, Walker AM (2018) Exploring the concept and experience of hope—empirical findings and the virtuous circle of hope. In: Krafft AM, Perrig-Chiello P, Walker AM (eds) Hope for a good life: results of the hope barometer international research program. Springer, pp 21–60. https://doi.org/10.1007/978-3-319-78470-0_1
Krafft AM, Guse T, Maree D (2021) Distinguishing perceived hope and dispositional optimism: theoretical foundations and empirical findings beyond future expectancies and cognition. J Well-Being Assess 4(3):217–243. https://doi.org/10.1007/s41543-020-00030-4
Krafft AM, Guse T, Slezackova A (2023a) Theoretical foundations and a transdisciplinary concept of hope. In: Krafft AM, Guse T, Slezackova A (eds) Hope across cultures—lessons from the international hope barometer. Springer, pp 23–44. https://doi.org/10.1007/978-3-031-24412-4_2
Krafft AM, Slezackova A, Marujo HÁ, Flores-Lucas V (2023b) Values and targets of hope. In: Krafft AM, Guse T, Slezackova A (eds) Hope across cultures—lessons from the international hope barometer. Springer, pp 55–102. https://doi.org/10.1007/978-3-031-24412-4_3
Kretz L (2013) Hope in environmental philosophy. J Agric Environ Ethics 26:925–944. https://doi.org/10.1007/s10806-012-9425-8
Lear J (2006) Radical hope: ethics in the face of cultural devastation. Harvard University Press. https://doi.org/10.4159/9780674040021
Lerner MJ (1980) The belief in a just world. In: Lerner MJ (ed) The belief in a just world. Perspectives in social psychology. Springer, pp 9–30. https://doi.org/10.1007/978-1-4899-0448-5_2
Lomas T, Hefferon K, Ivtzan I (2015) The LIFE model: a meta-theoretical conceptual map for applied positive psychology. J Happiness Stud 16:1347–1364. https://doi.org/10.1007/s10902-014-9563-y
Lueck MA (2007) Hope for a cause as cause for hope: the need for hope in environmental sociology. Am Sociol 38:250–261. https://doi.org/10.1007/s12108-007-9017-7
Maartensson H, Loi NM (2022) Exploring the relationships between risk perception, behavioural willingness, and constructive hope in pro-environmental behaviour. Environ Educ Res 28(4):600–613. https://doi.org/10.1080/13504622.2021.2015295
Marcel G (1951) Homo viator: introduction to a metaphysic of hope. St. Augustine's Press.
Martin A (2019) Interpersonal hope. In: Blöser C, Stahl T (eds) The moral psychology of hope. Rowman & Littlefield, pp 229–248
McDonald JE (2008) The spirit of hope and its near enemy indifference: a phenomenological continuum. In: Bauman W (ed) Hope: global interdisciplinary perspectives. Papers presented at the 3rd global conference hope: probing the boundaries. Inter-Disciplinary Press, pp 39–49
McGeer V (2004) The art of good hope. Ann Am Acad Pol Soc Sci 592(1):100–127. https://doi.org/10.1177/0002716203261781
McGeer V (2008) Trust, hope, and empowerment. Australas J Philos 86(2):237–254. https://doi.org/10.1080/00048400801886413

Meirav A (2009) The nature of hope. Ratio 22(2):216–233. https://doi.org/10.1111/j.1467-9329.2009.00427.x
Miceli M, Castelfranchi C (2010) Hope: the power of wish and possibility. Theory Psychol 20(2):251–276. https://doi.org/10.1177/0959354309354393
Miller JG (1999) Cultural psychology: implications for basic psychological theory. Psychol Sci 10:85–91. https://doi.org/10.1111/1467-9280.00113
Milona M (2019) Finding hope. Can J Philos 49(5):710–729. https://doi.org/10.1080/00455091.2018.1435612
Milona M (2020) Philosophy of hope. In: van den Heuvel S (ed) Historical and multidisciplinary perspectives on hope. Springer, pp 99–116. https://doi.org/10.1007/978-3-030-46489-9_6
Moellendorf D (2006) Hope as a political virtue. Philos Pap 35(3):413–433. https://doi.org/10.1080/05568640609485189
Moltmann J (2021) Theology of hope: for the 21st century. SCM Press
Naderi Farsani S, Abolghasemi MJ (2008) An investigation on the possibility of mutual understanding in a pluralistic world. In: Bauman W (ed) Hope: global interdisciplinary perspectives. Papers presented at the 3rd global conference hope: probing the boundaries. Inter-Disciplinary Press, pp 111–130
Nolt J (2010) Hope, self-transcendence and environmental philosophy. Inquiry, 53, 162–182.
Northcott MS (2020) Ecological hope. In: van den Heuvel S (ed) Historical and multidisciplinary perspectives on hope. Springer, pp 215–238. https://doi.org/10.1007/978-3-030-46489-9_12
Ojala M (2012) Hope and climate change: importance of hope for environmental engagement among young people. Environ Educ Res 18(5):625–542. https://doi.org/10.1080/13504622.2011.637157
Ojala M (2015) Hope in the face of climate change: associations with environmental engagement and student perceptions of teachers' emotion communication style and future orientation. J Environ Educ 46(3):133–148. https://doi.org/10.1080/00958964.2015.1021662
Olsman E (2020) Hope in health care: a synthesis of review studies. In: van den Heuvel S (ed) Historical and multidisciplinary perspectives on hope. Springer, pp 197–214. https://doi.org/10.1007/978-3-030-46489-9_11
Osmani NM (2008) Phenomenology of hope and despair: a Qur'anic perspective. In: Bauman W (ed) Hope: global interdisciplinary perspectives. Papers presented at the 3rd global conference hope: probing the boundaries. Inter-Disciplinary Press, pp 191–202
Peterson C, Seligman ME (2004) Character strengths and virtues: a handbook and classification. Oxford University Press
Pinsent A (2020) Hope as a virtue in the middle ages. In: van den Heuvel S (ed) Historical and multidisciplinary perspectives on hope. Springer, pp 47–60. https://doi.org/10.1007/978-3-030-46489-9_3
Rorty R (1998) Achieving our country. Harvard University Press
Rorty R (1999) Philosophy and social hope. Penguin Books
Rorty R (2002) Hope and the future. Peace Rev 14(2):149–155. https://doi.org/10.1080/104026502 20140166
Scioli A (2007) Hope and spirituality in the age of anxiety. In: Estes RJ (ed) Advancing quality of life in a turbulent world. Springer, pp 135–150. https://doi.org/10.1007/978-1-4020-5110-4_9
Scioli A, Biller H (2009) Hope in the age of anxiety. Oxford University Press
Scioli A, Chamberlin CM, Samor CM, Lapointe AB, Campbell TL, Macleod AR, McLenon J (1997) A prospective study of hope, optimism, and health. Psychol Rep 81(3):723–733. https://doi.org/10.2466/pr0.1997.81.3.723
Scioli A, Ricci M, Nyugen T, Scioli ER (2011) Hope: its nature and measurement. Psychol Relig Spiritual 3(2):78–97. https://doi.org/10.1037/a0020903
Shade P (2001) Habits of hope—a pragmatic theory. Vanderbilt University Press
Shade P (2019) Shame, hope, and the courage to transgress. In: Green R (ed) Theories of hope: exploring alternative affective dimensions of human experience. Rowman & Littlefield, pp 47–70

Shin SH, De Pretto L, Simpson A (2013) What does hope mean to people? The lay construct of hope under achievement-oriented settings in Australia, Italy, and South Korea. In: Karabin G (ed) Hope in all directions. Brill Publishers, pp 197–212. https://doi.org/10.1163/9781848882591_018

Singelis TM, Triandis HC, Bhawuk DP, Gelfand MJ (1995) Horizontal and vertical dimensions of individualism and collectivism: a theoretical and measurement refinement. Cross-Cult Res 29(3):240–275. https://doi.org/10.1177/106939719502900302

Slaughter RA (1993) The substantive knowledge base of futures studies. Futures 25(3):227–233. https://doi.org/10.1016/0016-3287(93)90134-F

Slezackova A, Krafft A (2017) Hope: a driving force of optimal human development. In: Mohan J, Sehgal M (eds) Idea of excellence: multiple perspectives. Panjab University, pp 1–12

Snow N (2019) Is hope a moral virtue? In: Blöser C, Stahl T (eds) The moral psychology of hope. Rowman & Littlefield, pp 171–187

Snyder CR (1994) The psychology of hope: you can get there from here. Simon & Schuster

Snyder CR (2000) Hypothesis: there is hope. In: Snyder CR (ed) Handbook of hope: theory, measures, and applications. Academic Press, pp 3–21. https://doi.org/10.1016/B978-012654050-5/50003-8

Snyder CR (2002) Hope theory: rainbows in the mind. Psychol Inq 13(4):249–275. https://doi.org/10.1207/S15327965PLI1304_01

Stahl T (2019) Political hope and cooperative community. In: Blöser C, Stahl T (eds) The moral psychology of hope. Rowman & Littlefield, pp 265–283

Stitzlein S (2019) Pragmatist hope. In: Blöser C, Stahl T (eds) The moral psychology of hope. Rowman & Littlefield, pp 93–112

Stockdale K, Blöser C, Stahl T (2019) Emotional hope. In: Blöser C, Stahl T (eds) The moral psychology of hope. Rowman & Littlefield, pp 28–44

Stotland E (1969) The psychology of hope. Jossey-Bass

Tennen H, Affleck G, Tennen R (2002) Clipped feathers: the theory and measurement of hope. Psychol Inq 13(4):311–317. http://www.jstor.org/stable/1448876

Triandis HC (1997) Cross-cultural perspectives on personality. In: Hogan R, Johnson J, Briggs S (eds) Handbook of personality psychology. Academic Press, pp 439–464. https://doi.org/10.1016/B978-012134645-4/50019-6

van Vliet J (2020) An ontology of human flourishing: economic development and epistemologies of faith, hope, and love. In: van den Heuvel S (ed) Historical and multidisciplinary perspectives on hope. Springer, pp 239–261. https://doi.org/10.1007/978-3-030-46489-9_13

Webb D (2007) Modes of hoping. Hist Hum Sci 20(3):65–83. https://doi.org/10.1177/0952695107079335

White KA (2009) Self-confidence: a concept analysis. Nurs Forum 44(2):103–114. https://doi.org/10.1111/j.1744-6198.2009.00133.x

Andreas M. Krafft Researcher and lecturer at the Institute for Systemic Management and Public Governance at the University of St. Gallen, Switzerland. His main research areas are the psychology of future thinking and transdisciplinary research on hope. For over ten years, he has been responsible for the international research network of the Hope Barometer.

Chapter 3
Horizons of Hope

A Philosophical Approach to the Practice(s) of Hope

Stefan Einsiedel and René Leibold

Abstract One of the oldest challenges in the history of philosophy is understanding the multifaceted nature of hope and identifying appropriate ways to deal with it. This introductory article takes a phenomenological approach, examining various philosophical and linguistic ideas and images used to describe human hope. The authors encourage readers to consider the terms 'hope' and 'horizon' in order to gain a deeper understanding of this vague yet alluring (and sometimes misleading) phenomenon. Both hope and horizon represent important yet intangible points of orientation, perceived differently by each person depending on their individual point of view. But is it fair to speak of one common horizon or one shared hope? The authors discuss how the individual 'I hope' can and should evolve into a broader 'we hope'. They use the Christian concept of hope as a concrete example of this process, not to determine whether Christian hope is 'well founded', but to explore what can be learned from religious practice for the social task of fostering a sense of shared hope in the face of the climate crisis. These different perspectives may gradually contribute to a deeper understanding of not only hope itself, but also of humankind as a species of constant hope- and horizon-seekers—just as Immanuel Kant approached the question 'What may I hope for?' by asking 'What is man?'.

Keywords Human hope · Phenomenological approach · Common horizon · Religious practice

S. Einsiedel (✉)
Center for Social and Development Studies, Munich School of Philosophy, Munich, Germany
e-mail: Stefan.Einsiedel@hfph.de

R. Leibold
Lassalle Institute, Zurich, Switzerland

Nordakademie, Elmshorn, Germany

T. Karcher SJ et al. (eds.), *Making Hope Possible*, Environmental Humanities: Transformation, Governance, Ethics, Law 7,
https://doi.org/10.1007/978-3-032-17719-3_3

3.1 Hope and Language as Human Characteristics

In the twentieth century, scientists gradually discovered that many animal species possess abilities that were previously thought to be uniquely human: not only the ability to use and produce tools for specific purposes, but also the capacity to selectively pass on information about these tools and other useful resources to other individuals, and the ability to communicate in ways that resemble language, sometimes even with the potential to differentiate between the past, present, and future (Andrews et al. 2024). What is apparently still uniquely human is the drive and need to constantly practice these abilities in order to further enhance and differentiate them. This situation is probably similar to another human phenomenon inextricably linked to the phenomena of 'language' and 'conceptualising the future': the ability (and need) to hope.

Ever since they started thinking about humanity, philosophers have speculated about hope, often viewing it as ambivalent: a human activity and quality, a temptation and a virtue, incomprehensible yet indispensable to human life. 'Dum spiro, spero' ('As long as I breathe, I hope') is a famous saying of the classic orator and philosopher Cicero (106–43 BC). To him and his contemporaries, hope was specifically human, whereas today we assume that it is typical, but not in all aspects exclusively, human. Perhaps it is accessible to all creatures who can contemplate their future. The initial feeling of hope could then be defined as a positive emotion that arises at the thought of a desirable and achievable future.

For Thomas Hobbes (1588–1679), hope was the dominant emotion that shaped human development and social life in many ways (Grethlein 2024). To him,

> hope is involved in the process of reasoning, is a reason for the emergence of religion, leads to competition in the state of nature [i.e. even to war] and at the same time helps to overcome it and create peace.

In *De Homine*, his contemplation on human nature, Hobbes implied that animals lack hope because they lack language, which for him is a requirement for hope—while for humans even the almost unimaginable 'can be hoped for, if it can only be expressed' (Hobbes 1658).

150 years later, hope also played a decisive role for Immanuel Kant (1724–1804): To him, hope and reason are both typically (and in his opinion: exclusively) human. Only mankind is capable of reason and hope and obliged to cultivate and spread both. In *Logic* (Kant 1800), the big question 'What is man?' is framed in the questions 'What can I know?' (for Kant a question of metaphysics), 'What should I do?' (for him a question of morality) and 'What may I hope for?' (for him a question of religion). But if the existence of God and eternal life cannot be proven by pure reason, why should we act morally? Wouldn't it sometimes be more logical to act immorally for our own benefit? To put it (almost too) simply: Ultimately, Kant is not sure whether he can *believe* in God—but he *must hope* with all his heart that there is a God (and that as many people as possible share this belief) so that it is logical to act morally; because we hope in a good and just God, it makes sense to be good and just ourselves (Grethlein 2024). No doubt, Kant's hope based in the sphere of

religion is exclusively human—as metaphysics seem to be exclusively human. His hope is not the kind of specific short-term hope that we experience when we hope to catch the bus in time or a lion hopes to catch a zebra. So, while we use the term 'hope' for so many different forms of expectations, would it help to differentiate at least between various intensities or spheres of hope?

To Ludwig Wittgenstein, who observed that animals 'could be imagined as being angry, fearful, sad, joyful, frightened but not as hopeful' (Wittgenstein 1963, 485), it seemed clear: only those who master the use of a language can truly hope. But could it not also be the other way round, that our language shapes (and maybe even masters) our hope? That would imply that we might hope differently in different languages. And that our hopes (and our idea of what hope is) might have become more and more differentiated as our languages evolved.

Just how differently hope can be interpreted in different languages can already be seen in the goddess of hope in ancient Greece and Rome. In Greek mythology *Elpis* was the personification of hope. She was the last spirit (or: daimon) to remain in the jar after Pandora unleashed the evils that were in it. The Greeks were not sure whether it was a blessing or a gift that Pandora did not release this last daimon: after all, deceptive, blinding hope might be the evilest of all evils. Etymologically, elpis stems from *elpizo* ('I anticipate'—meant in a neutral way) and hope was regarded by many classical philosophers as something highly dubious that should be approached with caution—you never knew what hope would lead you to. In contrast, the Roman goddess *Spes* (also the personification of hope, etymologically derived from *spera*, ray of light) was wildly popular. *Spero* ('I hope') meant looking, searching for something positive. No wonder that temples were dedicated to the Latin Spes and coins were minted, while the Greek Elpis led a shadowy existence. It was only in the Greek-language Bible that the word Elpis gained an optimistic note. We will come back later to how religious beliefs shape our understanding of hope, making just one observation for now: While in classical antiquity the distant future seemed less desirable to the (then dead) individual, in Christianity the future promised to be better than the present: hoping, in the sense of reaching out to God (who was the safe but distant future) was now seen as a gift and a virtue. This is also reflected in the modern words for hope which were coined at that time: *hoffen* (German for hoping) and the English *hoping* both indicate a positive, active, future-bound movement: 'a hop', an insecure but joyful leap, which, by the way, works only forwards and never backwards.

Many modern languages know the expression 'to ignite hope', which reminds us that hope, which in ancient Greek and Roman times was usually treated as something primarily individual, can be shared and passed on like a fire that warms and enlightens the entire neighborhood—we will also come back to this aspect of sharing later.

3.2 From 'I Hope' to 'I Hope, That'

Where and how does hope begin? Is it a vague—maybe way too vague—word for several different experiences? Can it be ignited by a natural process, like a fire, or does it have to be conceived by an intelligent being? For the phenomenologist Gabriel Marcel, hope is the 'answer from being' (Marcel 1957, p. 30) to the trials of suffering and despair. As part of the human condition, or the *conditio humana*, hope is born with every new baby and dies only when man dies, as the saying goes in many languages.

So, do we struggle with fear and respond with hope from the moment we are born? A baby's first breath is probably just a reflex, born not out of hope but perhaps accompanied by fear. But do we—hardly any later—hope for satiety when we are first hungry? We cannot hope for food because we do not yet know what it is. Is hope a basic human experience that begins right here? Does the practice of hope begin here? If so, this would mean that hope exists before language develops. It would also mean that we learn to hope through the fulfilment of our earliest desires. The experience of being satisfied by milk transforms the vague hope and/or desire to survive into the more specific hope of surviving by obtaining milk. We learn to focus our hope on something specific and good, and this continuous process paves the way for the habit of being hopeful. This suggests that, while hope itself may not yet be a virtue, the *attitude of seeking out worthwhile hopes* certainly is. This is the learning process of progressing from a vague 'I hope' to a more specific 'I hope that', as Marcel (1957, 30) describes the practice of hope.

And where exactly within this process does language break in? The moment I need to specify what I hope for, could that be the initial moment when language is born? And isn't language born out of hope for an answer? Some might say this is more desire than hope, but it seems more convincing that hope, desire and fear are all inseparably involved at the same time.

In this context, it might be helpful to consider Waldenfels' idea of responsivity (Waldenfels 2020): We are always answering to a world that gets back to us, confronts us, nudges us, reaches out to us as whole persons (not only as mind or as body but as the embodied and conscious person we are) and there is always an essential hiatus between every challenging experience of the world (which Waldenfels describes as our 'pathos') and our response. We are thus entangled in a two-step ('Zweitakt') system of pathos and response. Building on Waldenfels' idea, we could understand *hope also as a dual response* which starts right in this hiatus: first, hope is born as an emotional uprise against this challenging pathos ('I will not give in to that'), which in a second step enables (or matures into) a more concrete idea and/or action formed against this experience. Describing hope as our dual-nature response to the world we are part of might allow us to explain the wide spectrum of meanings that hope has for us as humans. In the complex interactive process in which my perception of myself (Heidegger's *Ich-selbst-sein)* is formed by my experience of being in the world (*In-der-Welt-sein*), hope is always the basis and part of each answer, and, yes, to some extent hope *is* the answer to the challenge of being human.

This directional character of hope (it directs my thoughts and actions into a helpful direction) is prerequisite for the virtuous character of hope. Hope can be both an aim and the act of aiming—but hope is not limited to that. In the next chapter, we will discuss the importance of moving from 'I hope' to 'we hope', from 'my hope' to 'our hope'. One reason for that is that it is quite difficult for a single person to define a 'really good aim' among the 'various options that are just a little better than the pathos I momentarily experience'. In our complex world, it often takes many eyes to define something as truly good.

3.3 From 'I Hope That' to 'We Hope That'

Hope might start as a fundamental personal experience (Marcel's 'I hope'), but like any other feeling, it cannot easily be shared and compared with others unless it is linked to something I hope for (Marcel 1957, 30; Held 2006). Hoping without a hoped-for remains without content and approaches an attitude of 'optimism', which, unlike hope, does not require a concrete hoped-for thing. I always judge my hoped-for, as is inherent in hope, as better than my current situation. Whether it is really better (or even objectively good) is *initially* phenomenologically irrelevant, I simply hope/assume it is. Hope in its early stage is very similar to two other feelings: optimism and longing. But hope can mature, while optimism stays a vague attitude that 'something good will happen'; and the personal feeling of longing continues to be a vague desire for something individually pleasant, which is not necessarily objectively good for me and for others.

Often, a maturing hope grows from an individual hope into a shared or collective hope—although this form of hope is sometimes more difficult to specify. It is so to say not an individual point a single person aims at, but a number of closely aligned points that define a shared 'hope space' (or 'horizon' as we will explain), a place full of good (individual and collective) options for the future. The 'I hope that', when aiming for the same goals grows into a 'we hope that'. There are many reasons why 'I hope' should become 'we hope', including the following: (1) Hope becomes clearer when it is shared. The good towards which hope (especially in the Kantian sense) should (and can be) directed is not always easy to see; it may require the collective and corrective effort of many eyes. Individuals can broaden their limited horizons by communicating with their fellow human beings, telling them what good things they see and learning from their perspectives. Like a candle lit in twilight, hope is also not an entirely private issue, it lights up the neighborhood and broadens the horizons of by-standers as well. As a result, (2) hope becomes bigger and more convincing when it is shared. A single candle cannot only enlighten its surroundings, it also radiates warmth which can ignite another light. As will be discussed later, this can imply the duty to 'hope and thus ignite hope in and for others'—when I share my small initial hope, others might also start hoping. (3) Hope becomes not only bigger, but also more attainable when it is shared. Hope always involves a certain personal uncertainty: will I ever achieve the hoped-for in my lifetime? Does it make sense to fight for climate protection or against a dictator if I can't be sure whether I will live to see the good or bad results? Aren't the chances of achieving the desired, good goal as an individual vanishingly small? Don't I perhaps even have a better chance of preserving my own life if I hide and don't resist, even if this further diminishes the hopes of many others? If I align or combine my personal hopes with a broader 'collective hope', I can foster a hope that goes beyond my own horizon of possibilities. Many significant hopes (such as finding a new cure against a deadly disease or winning the fight against climate change) cannot be achieved by most individuals who hope for them right now, but the collective of a hopeful humanity has the opportunity to do so ('I cannot reach my hope, but we can reach our hope'). Depending on how much 'I' experience myself as part of this 'we', I can live with the fact that 'we', but not 'I' will reach that precious goal. Generationally, I hope for others while enjoying a standard of living that my parents and grandparents could only hope for—and to some extent I enjoy it because they had hoped for it.[1]

Religious hope (which we will examine in the last chapter) is often also an 'I hope', embedded in a bigger 'we hope'. In the eyes of the believer, lighting a candle in an empty church or praying silently is not a resigned substitute for action, but strengthens an ongoing action. The believer hopes that this light does not shine in vain and trusts that this small symbol radiates hope—he or she always actively hopes for and with others.

3.4 Bold Hope Opens up New Horizons

Although collective hope may be more reliable and well-founded than individual hope, there is still a risk that it merely reflects a society's consensus, which could be outdated or based on wishful thinking. This is particularly true when collective hope assumes that current trends or the status quo will continue, or that unpopular reforms will not be necessary. In this instance, progressive individual hopes are required to guide collective hope in the right direction.

The example of Christopher Columbus, who literally opened 'new horizons' (although he never fully understood them), can help us to understand how individual and collective hope can influence and push each other. As we go through life, our world appears as a 'horizontal environment' in which the hoping subject seeks out and experiences different events intentionally and consciously. Our horizon is often the border between the *terra cognita* (the world we know) and the *terra incognita* (the parts of the world we don't know). It is the line where discoveries begin, not a

[1] From this generational perspective, it becomes clear that in the face of major intergenerational challenges, we have a duty to formulate well-founded hopes that extend beyond our personal horizons—and a duty to formulate them quickly and clearly enough, because hopes guide and enable decisive action. In addition, the generational perspective again reveals the difference between hoping and wishing. *Our hope* has to be well-founded, while *my wish* (which often pre-dates hope) can be anything I can imagine, however impossible or unrealistic it may be. Wishes are hypothetical, while hopes are those wishes that might (and should) come true someday.

defined border, but an imaginary frontier that moves back as we move forward. We process and understand these events (some of which are new discoveries) step by step, which in turn influences our future expectations and hopes. As this process of making sense of these occurrences takes time, we can differentiate between *horizons of events*, *horizons of experience*, and *horizons of expectation*. We can assume that new incidents are first understood pre-reflectively under our event horizons; under our experience horizons, we have processed these events-for example into learnings or narratives; a widening field of experiences empowers us to enlarge our expectation horizons (i.e. everything we can possibly hope for).

Hope pushes us towards the edge of our horizon of experience and beyond, towards the horizon of expectation. Hope in this sense doesn't mean a blind 'anything goes', but means a step (or a hop) from the world we know (and suffer from) to a wider world, which has more and better options. In this sense, a wide and well-founded 'horizon of expectation' enables confidence that hope is generally possible, and a specific hope is reasonable. These horizons of expectation stand in an indissoluble relationship to horizons of experience. It can be said that the expectation horizons 'emerge' from the experience horizons; they rise from the horizons of experience as drafts of a potential future. In this context, hope is experienced as a multi-dimensional emotion: as a characteristic attitude towards the world and an intuition that there is a concrete possibility of a better future. It is (again) both active and passive. Active in seizing the reins of action towards the better, passive as a mood towards the world and its horizon. Hope once again shows itself to be both a practice and a virtue, as a way (and art) of looking at the world. As Held put it (2006, 130f. with reference to Husserl):

> All occurrences of this life appear to us in horizons, and these in turn belong to the world as the horizon that is familiar to us non-representationally as the all-encompassing context of reference. The confidence that life will continue in a bearable way is therefore concretely anchored in a basic trust that the world understood in this way will continue to exist in the future.

In this sense, hope also manifests itself as a basic trust in the continued existence of a world that is good and worth living in. After all, hope is orientated towards changing things for the better (or preserving what is good). The simultaneity of active and passive parts of hope is also evident here. We never achieve what we hope for through our actions alone; a part, a surplus, always eludes our direct influence when we hope. If this was not the case, our actions alone would bring our plan to fruition, and we would not need hope, as determination alone would suffice.

Columbus hoped to discover a new sea route to India. We know he didn't succeed with *that*, but his journey illustrates how a hopeful attitude, based on experience, can open unexpected horizons. It also reminds us that there is a hope that motivates us to *start* a risky endeavour and a (different kind or aspect of) hope that helps us to *keep going*. Hoping for land, he knew: If no land comes into sight, the journey ends in disaster. Of course, fear was not the reason or the origin of this intensifying hope—it is just one of the many drivers of hopeful thinking. Columbus' hope was more than just hoping for a piece of land. And yet, the recurring contemplation of the horizon was also a growing 'hope for land'; a longing for a change, for no longer seeing only water, but something appearing on the horizon, showing, revealing itself. This hope is, as Klaus Held puts it, a 'hope based on the fulfilment of life through certain events that appear in the world.' (Held 2006, 126, translation by the authors).

It seems that all the different horizons of hope we just discussed have one thing in common. They all contribute to a basic human attribute, may it be individual or social: Not to stop searching and to hope for the even better to come. The *homo sperans* might be, facing the challenges of our times, our best chance to cope with a contradictional future, if we understand hope as a fundamental part of the *conditio humana* that makes us act towards things that are worth hoping for, overcoming despair. So, to hope seems the way we strive/live, or as Bloch (1980, 2) puts it:

> Primarily, everybody lives in the future, because they strive, past things only come later, and as yet genuine present is almost never there at all.

So, we live striving towards a future that might be better than the situation we are in. It is our striving in hope, our acting out of a hopeful emotion, which allows for new opportunities to rise on our personal horizons. And it is about much more than about our personal horizon(s), as Bloch (1980, 5) points out:

> Expectation, hope, intention towards possibility that has still not become: this is not only a basic feature of human consciousness, but, concretely corrected and grasped, a basic determination within objective reality as a whole.

3.5 The Need to Learn and Re-learn Hope

It is an art to hope widely and imaginatively and at the same time precisely and reasonably enough or as Bloch emphasized: 'It is a question of learning hope' (Bloch 1980, 1). In the face of the existential challenges of our time, can we afford to lack not only hope but also the act of hoping?

As previously discussed, the act (and art) of hoping that starts with our birth has to be practiced and refined in the way that it becomes a useful *hoping for*. Yes, hope is a matter of practicing, not only of experience, but also of experiencing and in that way a process of lifelong learning. But is that all what Bloch meant? We can gain a better understanding of his idea by examining the social hopes and the aims that we do hope and should hope for. Given that we are entangled in a world that is in danger due to our behaviour as human beings, we obviously need to improve our hope. We need to learn and relearn, shape and reshape our hopes.

Firstly, we must learn how our hopes correspond to the specific goals we aspire to—for ourselves and for others. This also means that we need to learn that there are things we should *not* hope for. When it comes to questions of sustainability and socio-ecological transformation, science, especially ethics, has a duty to talk much more about the hopes that must be prevented and ended. This involves dispelling false promises and illusions to make room for realistic hope.

Secondly, we should understand both learning and the lessons learned as strong imperatives to *act*, both as individuals and as part of society. If hope is followed by action, it fuels new hope. Coupled with action, the two-step-process of hope (which we discussed in relation to Waldenfels' *Zweitakt* concept of responsitivity) turns from a singular event into a repetitive process: so-to-say a two-stroke-engine that drives hopeful action. If we hope for the right things and if we understand our hope as an imperative act, we will survive or as Bloch said: '*Spero ergo ero,* because I hope, I will be' (Bloch 1968). And he iterated (1980, 1):

> The affect of hope goes out of itself, expands people instead of constricting them, cannot know enough of what makes them focused within, what may be allied to them by heart. The work of this affect requires people who actively throw themselves into the becoming to which they themselves belong.

But these hope-inspired actions have to be learned and trained. Only in this way—and with enough discipline—does hope become practice. Philipp Blom summarizes: (Blom 2024, 131–132):

> It is only through the application of discipline that this hope can be sustained, and only through this sustained effort can it gain a modicum of access to reality. [...] The mastery of any craft necessitates the acquisition of a technique that has been refined over an extended period, developing intricately between one's fingers and facilitating the transmission of this knowledge. The possibility of this active hope is contingent on the existence of connectedness, which, in turn, is facilitated by the aforementioned interconnectedness, transcending both temporal and social boundaries.

If we want to delve deeper into this discipline of learning to hope, both in the form of 'I hope' and 'we hope', religious practices of hope could provide valuable inspiration. Of course, we cannot judge whether the 'counterpart' or 'final goal' of religious hope truly exists, but for the time being we can define the term 'religious hope' as a long-distance-hope paired with an emotional conviction that the hoped-for is already partly hooked within my (the believer's) heart, where I can forefeel it as the best and most eternal counterpart a living creature could possibly have. Religious practice fosters a growing connection by which both I and this best goal draw me towards it. However, if this goal does not have the potential to truly call and draw me towards it, or if it dies when we (all humans) die, this hope, though still religious, is just a dream.

As previously mentioned, religious hope (which is first of all 'real' in the heart of each individual believer) can only be witnessed, described and developed if it remains an individual 'I hope' that becomes increasingly embedded within a larger 'we hope'. This implies that religious hope should (and indeed can) never be completely passive but always has (and normatively has to have) an active component. This may start with the small step—or better: drop—of sharing a hope,[2] whereby the individual believer lets go of at least part of his or her egocentricity and thereby joins and strengthens a larger stream of hope. A cold, static hope that is not somehow eager to join something larger cannot be called hope in a religious sense.

[2] Every prayer can be understood (in multiple ways) as an act of sharing, every religious act as an attempt to join (or at least to reach out to) something much bigger and better than oneself.

In the final paragraph, we will examine a particularly interesting 'stream of hope': during the pontificate of Pope Francis (2013–2025), the question of hope played a significant role in view of the social and ecological crises of our time. Pope Francis' spirituality has strengthened many of the Church's environmental initiatives and has also significantly shaped the conference from which this anthology emerged. Can we draw any inspiration from this form of religious hope to foster stronger hope in the face of the environmental and social challenges of our times?

3.6 A Few Practical Learnings from Religious Hope

In recent years, a number of academics have highlighted the interesting parallels between Pope Francis' political spirituality and the the *Capability Approach,* which was developed by the philosophers (and Nobel prize winner) Amartya Sen and Martha Nussbaum (e.g. Winkler 2016; Einsiedel 2021). All three of these thinkers demand social and ecological justice, but as a starting point they postulate to bring every person to a *minimum threshold of justice.* They therefore recommend realizing the right of all people (present and future generations) to a good life by first guaranteeing everyone at least the necessary 'minimum level of opportunities' (or capabilities) that enable further individual and collective growth/development.

Like Sen and Nussbaum, Pope Francis emphasizes that no form of knowledge should be overlooked in the fight against poverty and climate change and that art, especially literature, music and painting, can provide important inspiration and a source for new hope and thus new solutions. Pope Francis' love of art and expressive symbols was evident when he commissioned a new *porta santa,* a large bronze door, for the Jubilee in 2025. Traditionally, these Holy Doors are opened every 25 years in the four main papal churches in Rome to invite pilgrims from all over the world to symbolically pass *thresholds of hope.* For the 2025 Jubilee (whose motto was 'Pilgrims of hope'), Pope Francis had an additional Holy Door installed in the Rebibbia Prison in Rome to draw his flock's attention to a frequently overlooked aspect of Christian hope: even if there are thresholds of hope which most of us *cannot* cross, *we still have to go there*—especially for and with each other. Christians have the duty to go to seemingly hopeless places in order to look beyond their previous horizons; they must actively look for new hope and share it freely, to expand each other's horizons of hope. Pope Francis' approach to putting Christian social teaching into practice is typified by this urge to bring (or even to carry) every person to 'thresholds of hope': as *perfect* justice for *all* people is an important but elusive aim, a 'final hope' we all should share, it is all the more important to guarantee *sufficient* levels of concrete reasons for hope to *every person today.* It is important to note that the goal here is not to bring *religious* hope to everyone, but rather that religious hope, if rightly understood, should urge every believer to ensure that everyone else can also have sufficient hopes to have a 'good life'.

Of course, everyone has their own thresholds, but there are good reasons (including those discussed in the Sect. 3.3 titled 'From I Hope That' to 'We Hope That') to believe that most thresholds are relevant to more than one person. If there is a moral obligation to help everyone to reach a threshold of hope, this usually implies an obligation to bring different people together. The concept of thresholds of justice can even be used to derive a *Human Right to Sufficient Hope*. Historically, the right to 'the pursuit of happiness' (rather than a direct right to happiness itself), as proclaimed in the American Declaration of Independence, could be seen as an early predecessor of such a right to sufficient hope: While we are not entitled to everything we hope for, there is already happiness to be found in the pursuit itself—especially if this pursuit is not a simple pastime or for pure profit-maximization, but sense-oriented. We will soon come to this important shift from 'useful' to 'meaningful'.

But what does it mean to bring everyone to 'thresholds of hope', to places where people can begin to hope for themselves and/or others? These thresholds open up concrete 'corridors of hope'; they don't lead to utopian places but show concrete spaces of future possibilities similar to Robert Musil's *Möglichkeitsräume* (Musil 1978, 16). While these spaces may be unattainable for me as an individual, they must be *foreseeable from these thresholds* in order to be actually *attainable for 'us' in the future*. Some of the contributions to this conference booklet describe precisely these spaces of possibility and hopefully they will move more people towards these thresholds of new-found(ed) hope.

At this threshold, where concrete options become visible, we might begin to understand the dictum that 'hope has two daughters: anger and courage' (attributed to St. Augustine of Hippo, although it is more likely to have originated later). By better seeing and understanding what is concretely possible, hope may turn into (or at least foster) anger and/or courage. This is partly due to the fact that from this threshold it becomes clear whether hope is justified and whether efforts make any sense—prior to this, this was merely a vague feeling of optimism with no directional motion.

In Christian hope (and certainly also in many other forms of religious hope), 'we hope' means that 'we humans' hope *for* and *with* a good God.[3] In the eyes of a believer, an action embedded in this shared hope can make *sense*, even if it brings no individual *benefit*. Of course, this vastly changes, and usually broadens, the believers' horizon of options.

By shifting the focus from 'merely useful' to 'meaningful', Christian hope can be an act of liberation; but it also usually includes the mandate to move on. The Christian practice of faith finds meaning in the most unexpected places and invites us to consciously seek out these places, to pause there, but not to linger forever. In this way, even the cross, a sign of an absolutely 'senseless' violent death, a place of shattered hopes, could become a symbol of a new, greater hope. At its deepest level, Christian hope is often a hope found after suffering hopelessness. S. Žižek's long engagement with 'Christian atheism' led him to emphasize how important it is

[3] This begs the question: Can God *hope* for something, does the verb fit for someone who is omniscient? On the other hand, *hoping* could be the perfect form of loving self-restraint of an almighty being. Religious practice, especially prayer, would be an immersion in God's hope for this world, a hope that urges us to take action and in which we are not alone.

to go directly to and through the 'zero point of hopelessness'. In his book *Courage to Hopelessness*, however, it is not primarily newly won hope that is the prerequisite for decisive action, but lost false hope (Žižek 2018, 11).

This is not to say that suffering itself is good, but there is a strong Christian hope that it can make sense not to shy away from suffering, and to carry others through it as well: a kind of spiritual extension of Churchill's 'If you are going through hell, keep going'. Christians hope that they are not the only ones who move in such a situation. In many parables that Jesus tells about God, the Good Father/Shepherd/Landowner is also on the move, he actively seeks and meets those who are looking for him. In the religious sense, 'to hope' always implies movement: the hoped-for also hopes and moves (for and with) those who hope.

In this sense, religiously connoted hope not only encourages believers to 'carry on' but also to reconsider and change their direction. Rather than blind 'progress optimism', which can be a catastrophe in itself (because we assume things will always continue as before), a radical change of direction is possible and indeed welcome. Such a non-egoistic long-term attitude of hope makes it seem sensible (even if it is not personally beneficial), to go to great lengths to protect the climate, even if the goal is barely achievable. It enables people to see reversal not as failure of their previous lifestyle but as new hope.

Unfortunately, the positive aspects of religious hope described here are repeatedly used to justify religion itself by emphasizing its positive effects on society. This falls short and does not even come close to doing justice to any major religion or the God it hopes for. Rather than serving to justify religion itself, our discussion about religious hope should serve anyone, believer or not, who wants to hope in the face of the multiple crises of our time: it is about opening up and broadening our personal conception of what hope is, developing a culture of hope that supports more sharing and multi-dimensional conversion instead of uni-directional acceleration. This willingness to change direction and openness to be surprised is indeed deeply 'human'. Artificial intelligence learns from past data, continues to write trends, can calculate tipping points more precisely (the particular advantage of AI lies in the virtual testing of millions of options), but it lacks the imagination to think something 'completely new', to be surprised. Hope is indeed a human virtue and animals are closer to us in that respect than artificial intelligence is.

This chapter of the book is about what hope is and can be. The subsequent chapters primarily explore what gives us hope. The whole book is about becoming 'hopeful' to the right degree. Our future does not yet exist, it is being developed through hoping and planning and language is the primary tool that enables us to work on it together. The act of seeking, sharing, and improving our idea of what we hope for is a virtuous process of 'self-empowerment that is never complete and never meant to be' (Blom 2024, 132). In the end, it allows for the small but significant step (or 'hop') from 'I have a dream' to 'we share a hope'. In this process, the Latin *Dum spiro, spero* allows for a 'Because I hope, we will continue to live/breath': *Quia spero, spirabimus.*

References

Andrews K, Birch J, Sebo J, Sims T (2024) Background to the New York declaration on animal consciousness. Retrieved 11 Nov 2024 from http://www.nydeclaration.com

Bloch E (1968) Atheismus und Christentum. Suhrkamp

Bloch E (1980) Das Prinzip Hoffnung. Suhrkamp

Blom P (2024) Hoffnung: Über ein kluges Verhältnis zur Welt. Carl Hanser

Einsiedel S (2021) Das Potential der Armutsperspektive. Der „Blick von den Rändern" im Werk von Sen, Nussbaum und Papst Franziskus. In: Heimbach-Steins M (ed) Jahrbuch Christliche Sozialwissenschaften, vol 62. Aschendorff Verlag, pp 323–340

Grethlein J (2024) Hoffnung. CHBeck

Held K (2006) Idee einer Phänomenologie der Hoffnung. In: Lohmann D, Fonfara D (eds) Interdisziplinäre Perspektiven der Phänomenologie, vol 177. Springer, pp 126–141

Hobbes T (1978) Man and citizen. In: Gert B (ed) English translation of Thomas Hobbe's De Homine. Doubleday, pp 33–85 (Original work published 1658)

Kant I (1968) Logik, Physische Geographie, Pädagogik. Akademie Text-Ausgabe, vol 9. De Gruyter (Original work published 1800)

Marcel G (1957) Philosophie der Hoffnung. Paul List Verlag

Musil R (1978) Der Mann ohne Eigenschaften. Rowohlt (Original work published 1943)

Waldenfels B (2020) Responsivity and co-responsivity from a phenomenological point of view. Stud Phaenomenologica XX:341–355

Winkler K (2016) Befähigungssemantiken in der theologischen Ethik. Kritische Analyse der Rezeption des Capabilities Approach in der deutschsprachigen christlichen Sozialethik. In: Heimbach-Steins M (ed) Jahrbuch für Christliche Sozialwissenschaften, vol 57. Aschendorff Verlag, pp 269–294

Wittgenstein L (1963) Philosophische Untersuchungen. In: Schriften Band 1, Tractatus logico-philosophicus, Tagebücher 1914–1916. Suhrkamp, pp 279–544, as cited in Grethlein L (2024) Hoffnung. p 11

Žižek S (2018) Der Mut zur Hoffnungslosigkeit. S. Fischer

Stefan Einsiedel Lecturer and Managing Director of the Centre for Social and Development Studies (Zentrum für Globale Fragen) at the Munich School of Philosophy, Germany. Trained as a biologist and economist, his research focuses on the ethics of sustainability, socio-ecological transformation, climate justice, and global participation. Drawing on interdisciplinary perspectives from environmental ethics, development studies, and the Capability Approach of Amartya Sen, his work explores pathways towards sustainable and equitable futures.

René Leibold Entrepreneur and scholar with a PhD in Philosophy. His research focuses on entrepreneurship and narratives. He is a research fellow at the Lassalle-Institute and Lecturer at the Nordakademie. His business activities are primarily in the pet care & food industry and the human resources sector.

Chapter 4
Hope is a Matter of Time

Storytelling at the Margins of Extinction

Kevin FitzGerald and Charles Henry

Abstract This contribution explores the epistemological and communicative challenges of climate change, arguing that the fragmented nature of today's knowledge and storytelling systems undermines the temporal continuity and cognitive structures needed to foster environmental justice and hope. Considering the great difficulty of developing and managing a collective response to mitigate the deleterious effects of climate change, this chapter promotes the adoption of a virtual, digital library and workspace, framed by the principles of the Universal Declaration of Human Rights and the aspirations described in the Laudato Si' encyclical. Drawing on cognitive narratology, the project proposes a narrative framework grounded in integral ecology, intergenerational learning, and digital storytelling to address the shortcomings of academic, media, and technological infrastructures. The chapter identifies the paradoxical roots of the climate crisis, showing how human progress has led both to societal advancement and ecological collapse, with the greatest burden on those least responsible. Hope is conceptualized as both a psychological and narrative necessity, linking it to temporal awareness and human agency. Through a thought experiment applying narratology to climate discourse, systemic barriers to hopeful narratives across social media, news, and academia are highlighted. Building on this, the chapter examines how fragmentation in social networks, sensationalism in news, and academic specialization inhibit collective imagination and response. It further outlines the Re:Mind project—a global digital ecology for cultural preservation and intergenerational storytelling—as a means to foster empathy, connection, and equitable climate action. Finally, it is argued that while narrative cannot "solve" climate

K. FitzGerald (✉)
College of Design, North Carolina State University, Raleigh, NC, USA
e-mail: kmfitzg4@ncsu.edu

C. Henry
Council On Library and Information Resources (CLIR), Alexandria, VA, USA

American Academy of Arts and Sciences, Cambridge, MA, USA

C. Henry
e-mail: chenry@clir.org

T. Karcher SJ et al. (eds.), *Making Hope Possible*, Environmental Humanities: Transformation, Governance, Ethics, Law 7,
https://doi.org/10.1007/978-3-032-17719-3_4

change, rethinking our relationship to knowledge, time, and each other can catalyze transformation and generative sustainability.

Keywords Climate change · Digital library · Generative sustainability · Narrative of hope

4.1 A Composite Paradox

Responding to climate change and its predicted, and already manifesting, devastation is an unprecedented challenge. One aspect of this challenge is its scale. A typical description of climate change posits an accretion of carbon in the atmosphere, in the form of Greenhouse Gases (GHG), that began with the industrial revolution, adumbrating over the last 200 years reaching a pivotal point now that has triggered rising temperatures, often violent fluctuations in weather patterns, upsurging seas, wildfires, desertification, and infestations that threaten our sustained habitation of planet earth (US Department of National Intelligence 2021). Eventually every species and region of the planet will be touched by and denigrated by this encompassing phenomenon (Balsari et al. 2020; Feulner 2015).

Another element of complexity is that, paradoxically, the source of the catastrophe underway is advanced, cultivated, and highly rewarded human achievement. Earth has been terraformed by our visions, formulations, algorithms, and ideas, by our values and aspirations (Wallace-Wells 2019). The planet is an exhibit of the human mind made manifest. We have thought this world into being. The world today reflects human progress that has lifted many regions out of poverty, prolonged human life expectancy significantly, raised the level of education (again, in selected regions), and enriched the nutrients and array of food available for our sustenance in unprecedented variety. And now our innovative machinations conduce to an interlocking cascade of phenomena that have globally coalesced to scorch, drown, and asphyxiate us.

Further, the early phases of climate's roil are from a human societal perspective inequitable, biased, and unjust: those parts of the globe that have contributed the least to the industrialization and capital investments that aggregate to the climatological destruction are the first and foremost recipients of sweeping damage and existential privation (Balsari et al. 2020).

4.2 Hope as a Necessary Aspect of Our Response

To effect a different outcome of our predictive decline, hope is required to honestly and rigorously adjust our idea of human agency and place in the world. Hope is defined here based upon the research of Charles Snyder, clinical psychologist. In contrast to an emotion, hope is a cognitive trait centered on the pursuit of goals and built on two components: a sense of agency in achieving a goal, and a perceived ability to create pathways to achieve that goal (Barone et al. 1997).

As a state of mind, the affordances of hope can include the formation of memories that accumulate successful examples and instances of positive responses to climate change, formative decision-making that similarly accrue to building a larger framework of actions, and memories that assist in the reimagination of currently deleterious scenarios to a future less destructive, more reciprocally balanced between humanity and our home.

Another way to formulate the instantiation of hope is to define it as a narrative—a new story that can counter the various impeding stories we are continually affronted with. This new narrative should confirm our unique intelligence as the source of a possible sixth extinction and recalculate the satisfying physical and metaphysical metrics we use to celebrate ourselves; address the tragic inequality of effects of climate change on those who have fewer resources, support systems, and recourse to mitigate catastrophe; and describe a future that is more equitable, stable, and sustained through a new interdependency of humanity and the earth (Kolbert 2014).

As with any narrative, the transformative effects of hope require a belief in a temporal continuity: drawing upon experience and knowledgeable insight (that occurred in the past) selectively reconstituted as a desired state (performed in the present) that can then be imaginatively applied to a subsequent reality (future). Belief in a progressive fourth dimension that is subject to our intervention is a psychological condition necessary for the praxis of hope.

4.3 The Thought Experiment

The research we undertook included a thought experiment that helped to schematize the challenges a hopeful narrative will confront in our contemporary world: distilling the qualities and characteristics of other narrative types in a conscious mental realm of intensely competing storytelling. Our hopeful response to, perhaps mitigation of, the ruinous, socially, and economically staggered effects of climate change must contend with a wide array of other narratives that are duplicitous, fragmented, purposely distracting, and devoid of context and useful content. The thought experiment is essentially an application of cognitive narratology to help determine how we perceive ourselves in time and space relevant to the conception of hope.

How difficult can it be to make hope possible? Exceedingly. The necessary perceptual conditions for hope are systemically undermined in the twenty-first century by technologies and industries peculiar to our era. The prevailing narrative expressions of our interactions with each other, our acquisition and interpretation of current events, and our knowledge of the natural world are consistently fragmented, short term, tribal, and reductive. This is not a reference to disinformation, conspiracy mongering, or manipulated doubt and denial; rather, to the consequence of social media, authenticated news sources, and peer reviewed academic information. These aggregated, fractalized epistemologies impede a concerted response to our existential threat, thwart the application of hope, and the redress of environmental social justice.

In sum, the current methods of organizing, presenting, and formalizing information across popular and academic cultures is irreconcilable with the complex, interdisciplinary, and causal reasoning required to assess and address the existential risk of climate change, consequently impeding hope and the instantiation of environmental justice (McKibben 2014; Pinker 2014).

4.4 Three Examples Impeding Any Narrative of Hope

There are many instances of recurrent narrative structures that impede the belief of temporal continuity and the formation of a hoped-for response to our environmental crisis. Three seem especially pervasive and pertinent.

Social media: the problems with these networks are well established and documented extensively (Burton 2022; Carbonell and Panova 2017). These media encourage tribal formations, are reductive in simplicity of social engagement, and in some cases are themselves ephemeral, images vanishing within a fairly short time frame without a trace built upon very short snippets of information exchange, the brevity of expression promotes short-term thinking and removes context from many if not most exchanges. A reliance on singular images to express thoughts, anonymity, and a reliance on performative statements as opposed to a dialogue or substantive conversation epitomizes a kind of denaturing of nuanced, evolving thought formation. Self-referential environments of narrowly targeted "recommend" algorithms, and the idealization of body types and acceptable ideas, sublimate time to an eternal present.

The damaging causal impact of these media are also increasingly understood. Many people who use these technologies wish they could more easily cease; an addictive aspect of submitting to the social interfaces is often noted, as is the measured rise in anxiety and the attendant rise in instances of youth suicide· (Memon et al. 2018). The latter is of particular note, as self-immolation is often equated to a complete loss of hope, or an inability to imagine a more hopeful outcome.

A second incessant source of fragmentation and disconnect is the tremendously profitable and seemingly ubiquitous broadcast of news and current events. The methods of conveyance of the news have long been subject to inquiry and consternation for decades. Advertisements, social media, 24-h coverage of often repeated headlines, the endless unspooling of chyrons—think pop-up news feeds, breathless breaking news, floating pictures, and an unrelenting reliance on the cheap, reductive drama of contending forces and antithesis (often manufactured to gain attention and emotional connection)—foreclose the causal framework that surrounds us (Robertson et al. 2023). Fragments, shorthand, and snippets disassociate our world, not explicate it, and functionally obscure the depth and complexity required to understand our present danger.

"In the context of no context" describes not just partisan newscasts but more mundane and anodyne presentations. Many programs are less news than entertainment (the distinction has been blurring for decades). Most stories tagged as "human interest" are anomalous bits of a happy episode—a pet rescued, a lost child found, an act of kindness shared—while the more pressing stories of human interest—including our potential extinction—get almost no airtime (Trow 1997).

A third example of storytelling that impedes the development of larger contexts and deeper interconnections is, somewhat surprisingly, the traditions of storytelling in higher education. While social media and broadcast news can be categorized as popular culture whose negative influence is more easily intuited, the gravity of advanced education and its research are almost never so characterized. Universities are often considered a bulwark against the fragmentation evident in social media and contemporary coverage of current events, but this is misleading. We felt that there are enough similarities of narrative construction that warrant elucidation (Singh Chawla 2020).

Much of the problem with higher education can be traced to what we call the arithmetic of prestige (Grove 2022). Higher education is idiosyncratic and competitive, with institutions resolutely isolated. Rankings promote the distinctions and the competition for students, funding, faculty, and resources, impeding collaboration and cross institutional cooperation. Disciplines sport epistemologies that support their respective grammars, vocabularies, research methods, and conventions, hindering interdisciplinary and transdisciplinary research that is essential to address the open, dynamic, complex and networked problems we are facing (Dorst 2015). Professional advancement is often dependent on mastering the specialized wordplay, further proroguing collective action.

Compounding this scattering of knowledge is the high cost of research and the firewalls against accessibility that major publishers insist on. Not only is most disciplinary discovery tightly targeted to a small audience of readers, it is also at times impossible to access without significant institutional or personal funding. A challenge of the Lassalle-Institut conference was to develop methods of attenuating a looming disaster that would accelerate an effective and lasting set of solutions that would employ "a holistic and diverse field of transdisciplinary work such as natural, social, political, economic, technical scientific research, policy planning and implementation, as well as public dialogue and community communications." This is an intuitive and uncontroversial aspiration, but in reality, the transdisciplinary approach is made more difficult by the culture, structure, and reward systems of higher education itself.

Responding effectively to climate change demands explanatory and interpretative narratives of exceptional complexity that capture a rigorous, honest explication of events and their causal environmental cascades that are essential to grasp an integral ecology. Hope, as well as social justice, is similarly complex, instantiated over a longer term, and often subject to sophisticated dialogic development. These stories require time and context. The current cognitive narratological aspects of three major aspects of human communication are tragically inadequate, diminishing our agency explicitly or inadvertently to influence our future and effect a more positive antiphon to our calamity foretold.

4.5 Responses in a New Key

Fresh narrative formats and storytelling techniques can serve as an essential tool to modify the effects of climate change. "Tool" conforms to the general definition of an object that extends our ability to modify features of the surrounding environment and, in this instance, modify ourselves. The severe cognitive constraints imposed by our splintered information landscape beg us to ponder the converse of its constraining affordances. If we do construct benevolent stories of hope to transmute our plight, the contemporary vessels of communication do not easily, if at all, permit the complexity and integral aspects of an anthropogenic ecological collapse to be accurately articulated. That message and those media do not fit.

Considering these deficiencies inherent in the prevalent modalities and venues of public human expression, we have concluded that working within existing frameworks of prominent knowledge configurations is difficult at best, and probably futile. In a figurative sense, we need to leave this world in order to save it.

4.5.1 The Turn to Digital

While the term "digital library" has been interrogated, and examples constructed, for over half a century, its potential has yet to be realized, in part because we seem to focus on using this technology to enhance traditional analogue procedures and methods. "Digital" makes an old card catalog faster, sleeker, and with access to a dizzying choice of resources. "Digital preservation" similarly evokes high-resolution surrogates of cultural objects that glow enticingly on our screens. Yet for both these terms a kind of stasis obtains: a virtual box still comes to mind with both constructs.

A new project, Re:Mind (CLIR 2025) has begun as a response to these existential conditions. Re:Mind, an advanced, regenerative digital library supports complex empirical research and also serves as a storytelling platform for analysis, modeling, and capacity to technically support multimedia narrative expression of voices from international regions of the planet, new stories and forms of expression that facilitate hope, foster social justice, and delineate effective responses to climate change, support intersubjective understanding on a global scale to attain purchase on the catastrophe that looms. A rigorously interactive and reciprocal digital environment, Re:Mind serves as a promissory note to generations coming of age and to come.

The project builds from a proven technical stack that federates almost any metadata, regardless of schema. It also employs AI against datasets that are trustworthy and peer reviewed (hence preventing hallucinations) while also utilizing Retrieval-Augmented Generations (RAGs) to further refine the AI analysis. It brings together living social networks—actual regional and institutional communities of students, teachers, and other independent and curious individuals—through transnational curriculum development, comprehensive training, exchange opportunities, fellowships, and iterative sharing of new discoveries.

Its subtitle is "connecting life to knowledge during climate destruction," emphasizing the capacity-building programs it hosts for communities that collect, document, preserve, and share unique cultural heritage before it is lost. The beliefs, values, and behaviors captured in those artifacts and traditions can then be applied to build resilience in the face of climate change.

4.5.2 *Project Structure*

The identification and prioritization of cultural resources to be digitized will be done by local residents; the project is solely at the discretion of the population served by this heritage. The digitization will be accomplished largely by younger generations using smart phones. The surrogate objects captured will be processed as if belonging to a single, worldwide digital library, with simple metadata applied, in most cases automatically. The metadata will be harvested and made accessible to search the Re:Mind database without cost. Teams around the world will build tools, applications, and augmentations to the general platform to aid discovery and reuse. A critical element in the project is the adaptation of these resources for teaching curricula and research worldwide.

What would otherwise be irretrievably lost will be secured and given voice. The resources so preserved will include humanistic expressions as well as rare and obscured scientific information that it is expected to enrich our understanding of climate change and the complex causal cascades it entails.

The first phase of Re:Mind focuses on Africa, where a coalition of nearly 500 institutions and research centers (Association of African Universities, UbuntuNet) have signed MOUs to begin work. Other formal partners include the Reserche Francophone Numerique (RFN), UNESCO, International Centre for the Study of Preservation and Restoration of Cultural Property (ICCROM), and the Integral Ecology Research Network (IERN).

4.5.3 *Generative Sustainability*

Re:Mind entails rethinking of key terms associated with projects of this reach: infrastructure and sustainability. *Infrastructure* typically refers to build, physical components of systems that permit the operation of our society, and can include services, practices, tools, and procedures necessary for the ongoing maintenance of these systems. In a digital knowledge ecology, wires, cables, software, algorithms, and servers are fundamental infrastructure, along with rules of organization, professional support skills, standardized tools, and applications of inquiry and analysis.

Sustainability is generally defined in financial terms: durability correlate with the supply of ongoing funding, available post-grant, that prolongs the project without relying on continual investment by the original funding source. This emphasis is understandable, as it signals ongoing institutional commitment with the possibility of augmenting the project. It is, like infrastructure, reductive and formulaic. Without question, grant funding that insists on adequate infrastructure and ongoing financial outlays privilege those institutions and organizations that already enjoy a more robust network of physical systems and professional expertise, along with money to extend the project.

An expanded concept of infrastructure would comprise an applied interdependence of support across institutions, cultures, and generations, gaining an organic complexity and motion to infrastructure that is often missing. A less bounded, more thoughtful approach would attract new collaborators and contributors who know the project is not short lived. The technological platform must similarly evolve; a longer time frame also can attract more of a community of technologists who can work together to mutual benefit that would be more difficult in a shorter build out: team sourcing and alliance maturation. Other benefits of achieving interdependence include integrating the knowledge and resources developed by longer term projects into teaching and research; curricular development and innovative practices of reuse and repurposing of this knowledge would be encouraged. Architecting platforms that welcome new practitioners, new resources, and fresh perspectives would be integral to the design.

Interweaving sustainability and infrastructure as symbiotic and undetachable functions would provide opportunity for larger scale community building, infrastructure development, and diverse revenue streams. The time and investment for greater systemic social and intellectual engagement repurposes these efforts to become less of a project in the traditional sense, more of a catalyst for progressive entanglement. The more aligned the functions of sustainability with infrastructure, the more enduring the enterprise. We have named this more encompassing approach "generative sustainability," and it is a core feature, and requirement, for Re:Mind.

4.5.4 Attracting Younger Generations

We make a special effort to attract younger generations as key to the project's durability, and essential to achieve generative sustainability. Through our integrative approach, young people directly participate in safeguarding, sharing, and learning from cultural heritage threatened by climate's ruin. They in turn pass this knowledge to those who follow, creating a living network of discovery, purpose, and focused action: across generations, the abiding inheritance of our essential humanity.

The dimensions of the collateral effects of the project can be modeled as cognitive displacement, an effect of storytelling more colloquially termed "change of mind." Perceptions, definitions, approaches, responses, degree of awareness relating to climate change are altered by the project.

With perhaps tens of millions of young people working concertedly to save and reuse our cultural heritage, it is possible.

1. The general understanding of the urgency of climate change will be enhanced,
2. The costs and disruptive dimensions of climate change will be better appreciated,
3. The relation of climate change as a core challenge to humanity—recontextualizing what is currently a rather statistical set of propositions—could be better grasped,
4. Denying climate change as a phenomenon will be more difficult,
5. Counter arguments dismissing climate change will be less tenable and creditable,
6. The powerful effects of an SIS would become better appreciated,
7. More collective action in response to climate change could be catalyzed, and
8. Youth provide the foundation for the continuity of hope.

4.6 Conclusions

We do not turn away from the physical world but rather seek to build a transdisciplinary, interactive human/digital environment essential for empirical inquiry and the composition and preservation of narratives of hope and justice, architected by a technical platform that both receives, engages, and enlightens us collectively to better understand the systemic weave of life we have marred, and the venture needed to mend it.

While many programs and projects designed to ameliorate the effects of climate change are critically important, there is profound anxiety among younger people that these efforts are not adequate for the enormity of the crisis. They look for more comprehensive, structured engagement, preferably on a planetary scale. Policy making is also appreciated, as are the many accords and promises over the years to collaborate at a national level to lower carbon emissions and check the rise of temperatures. Laudable certainly, but these have not proven efficient or effective.

So it is time to build, to apply our commonwealth of knowledge, our beautifully arrayed cultural legacy, and our sophisticated technologies to foster a more rigorous engagement with our crisis and, if successful, a collective change of mind. That in turn can lead to more informed decisions and concerted action. It is estimated that 92 million selfies are taken each day around the world. The simple act of turning a phone's camera away from oneself and pointing it to a tangible instance of our shared culture and pressing click safeguards and sustains it by connecting us to a living, intergenerational community of practice and purpose that could be, with this slight roll of the wrist, enough to stir the natural concordance between us and our solitary home, the revelation of a self more encompassing of spirit, and salient of hope (Henry 2025).

References

Balsari S, Dresser C, Leaning J (2020) Climate change, migration, and civil strife. Curr Envir Health Rpt 7:404–414. https://doi.org/10.1007/s40572-020-00291-4

Barone D, Maddux J, Snyder C (1997) Social cognitive psychology: history and current domains. The springer series in social clinical psychology. https://doi.org/10.1007/978-1-4615-5843-9
Burton T (2022) The impermanence of social media. PBI: Pennsylvaia Bar Institute. Retrieved on 10 Jan 2025 from https://go.pbi.org/blog/the-impermanence-of-social-media
Carbonell X, Panova T (2017) A critical consideration of social networking sites' addiction potential. Addict Res Theory 25:48–57. https://doi.org/10.1080/16066359.2016.1197915
CLIR (2025) Re:Mind. Retrieved on 10 Jan 2025 from https://remind.clir.org/
Dorst K (2015) Frame innovation: create new thinking by design. MIT Press
Feulner G (2015) Global challenges: climate change. Glob Chall 1:5–6. https://doi.org/10.1002/gch2.1003
Grove J (2022) Journal prestige still matters. Inside Higher Ed. Retrieved 10 Jan 2025 from https://www.insidehighered.com/news/2022/08/04/journal-prestige-still-important-how-scholars-judge-one-another
Henry C (2025) Making the case for hope: the idea of a not for profit in a fractured time. Retrieved on 12 June 2025 from https://www.clir.org/2025/06/making-the-case-for-hope-the-idea-of-a-not-for-profit-in-a-fractured-time/
Kolbert E (2014) The sixth extinction: an unnatural history. Henry Holt and Company
McKibben B (2014) The age of missing information. Random House Trade Paperbacks, Westminster
Memon AM, Sharma SG, Mohite SS, Jain S (2018) The role of online social networking on deliberate self-harm and suicidality in adolescents: a systematized review of literature. Indian J Psychiatry 60:384–392. https://doi.org/10.4103/psychiatry.IndianJPsychiatry_414_17
Pinker S (2014) Why academics stink at writing. The chronicle of higher education, the chronicle review. Retrieved on 24 July 2025 from https://www.chronicle.com/article/why-academics-stink-at-writing/
Robertson CE, Pröllochs N, Schwarzenegger K, Pärnamets P, Van Bavel JJ, Feuerriegel S (2023) Negativity drives online news consumption. Nat Hum Behav 7:812–822. https://doi.org/10.1038/s41562-023-01538-4
Singh Chawla D (2020) Science is getting harder to read. Nature Index. Retrieved on 24 July 2025 from https://www.nature.com/nature-index/news/science-research-papers-getting-harder-to-read-acronyms-jargon
Trow GWS (1997) Within the context of no context. Atlantic Monthly Press, New York
US Department of National Intelligence (2021) Global trends 2040: a more contested world. Retrieved on 24 July 2025 from https://www.dni.gov/index.php/gt2040-home
Wallace-Wells D (2019) Vaclav Smil on the need to abandon growth. Retrieved on 5 July 2025 from https://nymag.com/intelligencer/2019/09/vaclav-smil-on-the-need-to-abandon-growth.html

Kevin FitzGerald Trustee of the Council on Library and Information Resources (CLIR) (USA) and liaison with the Integral Ecology Research Network (IERN). Kevin holds a Doctor of Design (DDes) from North Carolina State University College of Design (USA). His design practice and research address complex, networked, adaptive, and common pool problems.

Charles Henry President of the Council on Library and Information Resources (CLIR), USA. An elected Fellow of the American Academy of Arts and Sciences, he has published widely on digital humanities, leadership, and innovation in higher education administration. His current work explores the perceptual transformations and cognitive strategies required to address climate change, with particular attention to the cultural and intellectual dimensions of sustainability beyond purely technical and engineering solutions.

Part II
Local and Regional Case Studies: Evidence for Hopeful Perspectives on Sustainable Development and Environmental Justice

Chapter 5
Ecosystemic Transdisciplinary Action Research: A Methodology for Regenerativeness

Amar K. J. R. Nayak

Abstract Ecosystemic Transdisciplinary Action Research (ESTAR) is a theoretically coherent and practically balanced methodology for a holistic and systemic diagnosis and synthesis of regenerative solutions in different contexts, making hope possible for all. ESTAR argues that (a) Transdisciplinary Methodology needs to go beyond the anthropocentric perspective and adopt an ecosystemic perspective with ecology at the core of its systems analysis, and (b) Research Methodologies, including Transdisciplinarity (TD), need to clarify and adopt an axiology that is coherent across all stakeholders and entities in an ecosystem to make regenerativeness and hope possible for all. ESTAR's ontology, epistemology, logic, and axiology are rooted in AIESS (all interacting evolving systems science) and are based on the axioms of interconnections, interdependence, interactions, and weakest others-centered intentionality. It is a non-dualism based, all-interacting, and evolving paradigm of thinking, doing, and living. Problem-solving is the primary objective of this methodology, which leads to the creation of authentic and original knowledge. This chapter is presented in four sections. First, it explores the existing strands of TD and discusses the key features of each strand in a chronology, bringing out the evolutionary aspect of this knowledge system. With its ecosystemic perspective and the weakest other-centered axiology, ESTAR methodology advances the body of knowledge on the methodology of research and practice for making hope possible. Second, it explains ESTAR in detail, comparing the approaches of different research methodologies in social, economic, political, natural, and applied sciences. Third, it shares the findings of the application of ESTAR in resolving water, soil health, and surface air temperatures of a small farming system, leading to a safe, healthy, nutritious, profitable, and regenerative farming system. Fourth, by extending its AIESS Framework with the spectrum of design variables in agriculture across other critical dimensions of an ecosystem, it highlights the relevance of ESTAR in diagnosis and resolving the polycrisis in different systems and ecosystems, suggesting that it is possible to make hope for all.

A. K. J. R. Nayak (✉)
Professor of Strategy & Chairperson of AIESS Centre, Xavier Institute of Management, XIM University, Bhubaneswar, India
e-mail: amar@xim.edu.in

T. Karcher SJ et al. (eds.), *Making Hope Possible*, Environmental Humanities: Transformation, Governance, Ethics, Law 7,
https://doi.org/10.1007/978-3-032-17719-3_5

Keywords Transdisciplinarity · Systems Science · Interactions · Intentionality · Uncertainty · Non-dualism · Axiology · ESTAR · AIESS

5.1 Transdisciplinary Methodology: A Review

There have been different approaches or methodologies for diagnosing and analyzing issues. The disciplinary approach has been the predominant way to study various phenomena. Some popular research methodologies include historical, case, survey, grounded theory, and ethnography. These methodologies could be at the etic and emic levels of engagement of the researchers or based on the source data (ideational or materialistic), sample size (deductive or inductive), or the nature of the analytical or technical skills of the researchers (qualitative or quantitative). Etic refers to a position where the researcher sees the problem from a distance and has not been part of the context where the problem occurred. Emic refers to the researcher's position in which the researcher is embedded in the context of the problem. However, it has increasingly been felt that the long-held disciplinary approach to diagnosis and analysis has had little effect in resolving complex issues.

Some key assumptions of the above disciplinary approaches include reductionism, separability (dualism) of issues in a context, and non-contextual and temporal nature. Increasing awareness of the above limitations for knowledge to be relevant in resolving complex problems of the living world has led to increasing interest in multidisciplinary and interdisciplinary approaches. Health Care and Agroecology are areas where these approaches have been most adopted in recent years.

The multidisciplinary approach attempts to study a problem through the lens of multiple disciplines, covering as much ground as possible so that the issue's complexities are addressed. It assumes unity within the space over underlying coherence within the space or context of study.

As an improvement of the multidisciplinary approach, the interdisciplinary approach attempts to explore the interconnected issues of two or more disciplines. In other words, it tries to find coherence in disciplinary explanations of issues related to a given problem.

Multidisciplinary and interdisciplinary approaches advance traditional disciplinary approaches by appreciating the interconnections of knowledge systems based on disciplinary boundaries. It breaks away from the assumptions that knowledge is bounded within disciplinary boundaries. However, like the disciplinary approach, both multidisciplinary and interdisciplinary approaches are oriented and focused on knowledge creation. The social context, problem context, or reality is just a medium for creating knowledge using the above approaches. Most importantly, they are based on an underlying assumption that there is certainty of knowledge in nature; hence, once knowledge is created, it can be applied to solve the problems of reality. While this assumption may broadly apply in highly volatile, evolving, complex systems of reality, the above assumption may not be very helpful.

Further, the interdisciplinary approach has advanced to a transdisciplinary one to better explain the growing complexities in the economy, society, and environment. The transdisciplinary approach appreciates issues' non-separability (non-duality) as expressed in the Charter of Transdisciplinarity. The transdisciplinary approach, as we see in practice today, is unique in orienting research to a problem-centred methodology, and the advancement of a discipline is less concerned than the disciplinary approach.

Since the early seventies, the transdisciplinary approach has gradually evolved, acquiring a rich history of thinking and practice. From being holistic, education gradually became reductionist with industrialization and subsequent specialization. The measurability of data and empirical evidence gradually became the cornerstone of science. However, the reductionist explanation always fell short of the holistic nature of our reality. Therefore, the need for holistic education in the early seventies gave rise to this new transdisciplinary method of thinking and practice. As the method of thinking had already been disintegrated and focused on disciplines, the recombination process started with multidisciplinary approaches and interdisciplinary integration. Systems thinking became the basis for the above new recombination methods to think holistically. The latest wave of the transdisciplinary approach in the nineties focused more on problem-solving than knowledge creation. The philosophical and theological search for problem-solving also saw the value of a transdisciplinary approach. However, there has also been a stream of thinking that aims to figure out coherence in things that are happening at different levels of reality through a transdisciplinary approach. These six different categories of transdisciplinary thinking and practice that have evolved since the seventies are as follows:

5.1.1 *Holistic Knowledge for Effective Problem-Solving*

Against the backdrop of increased reductionism and abstraction by the various scientific fields, the need for a comprehensive understanding of reality for problem-solving competence in higher education was strongly felt in the seventies. It emphasized the need to re-integrate Knowledge across disciplines to be relevant to everyday practice.

Jantsch (1970, 1972) proposed integrative educational frameworks, especially at higher-level education that combined multiple disciplines, advocating for systemic solutions rooted in self-organization and holistic thinking. His approach laid the foundational principles for transdisciplinary research (TD). His ideas primarily focused on education and theoretical systems, with limited exploration of participatory processes or real-world applications.

Piaget (1972) highlighted the importance of interdisciplinary learning in fostering intellectual development and a deeper understanding of complex problems. He believed that integrating disciplinary knowledge could enrich holistic education and cognitive learning. While his work influenced early ideas about TD, it remained largely theoretical, focusing on education rather than addressing practical societal or ecological challenges. His contributions provided a conceptual framework at the early stages of the TD Approach.

5.1.2 Multidisciplinary and Interdisciplinary Approaches

These initiatives were part of the early stages of TD development. The efforts were to bridge disciplinary gaps and include diverse academic perspectives for practical, real-world applications. After the early seventies, the deliberations on Transdisciplinarity peaked in the early nineties, a momentum that has continued for over three decades.

Mittelstraß (1992, 1993) emphasized the concept of unity within TD, advocating for knowledge synthesis that transcends disciplinary boundaries. In his 1992 paper, he explored the epistemological basis of TD, linking it to the idea of overcoming fragmented scientific approaches. In 1993, he argued that TD fosters a unification of science, integrating its practical applications with theoretical coherence. His contributions positioned TD as a vehicle for creating coherent knowledge systems. However, its practical strategies for implementation were limited.

Rosenfield (1992) highlighted the potential of TD research in strengthening connections between health and social sciences. She emphasized that TD can sustain and extend the integration of diverse knowledge systems to address global health challenges. By focusing on actionable solutions, her work provided a practical lens for TD applications in the health sector. However, her scope remained limited to health-related contexts, leaving out broader societal or ecological considerations.

In *Science for the Post-Normal Age*, Funtowicz and Ravetz (1993) argue that traditional science is inadequate for addressing complex, high-stakes societal challenges. They propose "post-normal science," a paradigm where uncertainty, value disputes, and urgent decisions dominate. Post-normal science emphasizes extended peer communities, stakeholder inclusion, and pluralistic knowledge integration. This framework challenges traditional disciplinary boundaries, advocating for participatory approaches. While impactful, the approach assumes ideal stakeholder engagement, which can be challenging to achieve in practice.

The Charter of Transdisciplinarity (1994) formalized the principles of TD, emphasizing the integration of knowledge to address global challenges. It introduced universal principles advocating for moral responsibility in scientific research and the importance of integrating science, culture, and society. Among other scholars, the three prominent signatories included Edgar Morin, Basarab Nicolescu, and Lima de Freitas. The Charter was groundbreaking in defining the ethos of TD, which impacted its subsequent specific developments.

Meyer (2007) explored the role of framing in TD research, highlighting how stakeholder perspectives shape problem definitions and solutions. His work demonstrated how framing mediates collaboration in TD, enriching interdisciplinary efforts. Meyer's insights were valuable in showing the dynamic nature of problem-solving in TD projects by ensuring social accountability in the research outcomes.

Lawrence (2010) argued for combining multi-, inter-, and transdisciplinary approaches to tackle environmental and societal issues. He introduced epistemological pluralism as a method to integrate diverse knowledge systems. His work broadened TD's methodological scope, emphasizing collaboration and synthesis. However, Lawrence's focus remained theoretical, lacking concrete operational models to apply TD principles effectively in real-world settings.

Jahn et al. (2012) emphasized the need for reflexivity and mutual learning to address institutional barriers in TD research in the local context. They highlighted integrating scientific and societal knowledge as a key factor in successful TD projects. Their work provided important insights into collaboration and institutional challenges at local levels, but they could not integrate more complex ecological or global application spaces.

Another key proponent of integrating community stakeholders in TD, Nowotny et al. (2003), updated the concept of Mode-2 knowledge by emphasizing societal resilience and accountability in research. She linked TD's approaches to emerging ecological and social challenges, stressing the importance of participatory knowledge systems.

Focusing on integrating methodologies across scientific disciplines, Kötter and Balsiger (1999) analyzed the dynamic challenges of interdisciplinarity and transdisciplinarity. They argued that TD requires continual adaptation of scientific methods and emphasized the importance of collaboration between diverse fields. Their work advanced the operational understanding of TD but lacked detailed mechanisms for ensuring stakeholder involvement or addressing societal applications. While integrating multiple disciplines and stakeholders helps solve local problems, it cannot find synergy and coherence with more complex contexts, such as ecological or global contexts.

5.1.3 Philosophical and Theological Interpretations

Philosophical and Theological interpretations of transcendence have also attempted to facilitate transdisciplinary research. This approach of TD has been based on dialogue among stakeholders from different knowledge systems. Ananta Kumar Giri, Larisa Kiyashchenko, and Peter Osborne are a few scholars who contributed to this line of argument.

Giri (2002) introduced "creative transdisciplinarity," emphasizing transformative dialogues and relational approaches across disciplines. He critiqued rigid disciplinary silos, advocating for philosophical exploration and collaborative problem-solving. His contribution linked TD with relational thinking, fostering its transformative potential. Though it enlarged the perspective on including experts from different knowledge systems, it hardly mentioned the inclusion of lay persons from the problem context in solving the problem.

Kiyashchenko (2017) emphasized boundary-crossing approaches in TD, integrating scientific, practical, and religious perspectives. Her dialogical synthesis offered a novel method for addressing societal challenges. While her work broadened TD's epistemological scope, it placed disproportionate emphasis on spiritual dimensions, limiting its applicability in concrete problem-solving contexts.

Osborne (2015) critiqued disciplinary norms, advocating for transversal and dialectical thinking in TD. He argued that reflexivity and the breakdown of disciplinary silos are crucial for impactful TD research. Osborne's work strengthened TD's theoretical foundations but lacked detailed operational frameworks for real-world applications, limiting its immediate applicability.

5.1.4 TD Rooted in Systems Thinking

With an increased understanding of the interconnected and evolving nature of systems, the basis for complexities and wicked problems, the application of Systems Science and Systemism became intrinsic to TD research. Systems thinking was seen as a way to deal with more complex issues and contexts, as well as contradictions and complexities. So far, the three scholars who emphasized advancing TD include Basarab Nicolescu, Julie Thomas Klein, and Wolfgang Hofkirchner.

Nicolescu (2010) proposed the "Hidden Third," a concept that integrates multidimensional realities and reconciles contradictory perspectives. His philosophical approach enriched TD's epistemological foundations by linking scientific and philosophical insights. While groundbreaking in theoretical depth, Nicolescu's work lacked practical applications for solving concrete societal or ecological challenges, leaving it as a primarily conceptual contribution.

Klein (2014) linked TD with systems science, focusing on its capacity to address "wicked problems" through interdisciplinary strategies. She emphasized the importance of complexity theory in TD, providing a strong theoretical framework to integrate multiple disciplines, including social and ecological, to solve complex real-world problems. However, the practical application of integrating participatory aspects was not elaborated.

Hofkirchner (2017) proposed "systemism" as a TD framework for addressing global challenges. His approach integrated systems thinking with TD, offering a robust methodology for tackling complex societal problems. However, the theoretical emphasis on systems frameworks overshadowed the importance of participatory processes and stakeholder engagement in TD practice.

Drawing from the General Systems Theory, the above Systems thinking frameworks provided a tremendous theoretical foundation for disciplinary integration, operationalizing the role of participatory methods into this approach was unexplored, limiting its practical application in stakeholder-driven TD research.

5.1.5 Stakeholder Inclusion

Diverse stakeholders in the problem context must be included through participatory methods to enhance understanding of an evolving reality in the living world. Operationalizing the integration of stakeholders from the problem context with the domain experts has been a significant effort to understand and solve the problem better. M. Gibbons, Helga Nowotny, Hirsch Hadorn, Christian Pohl, and Rodrick Lawrence have significantly contributed to this line of development in the transdisciplinary approach.

Gibbons et al. (1994) introduced the concept of "Mode-2 knowledge production," emphasizing application-driven, context-sensitive, and participatory research. They argued that societal challenges require collaborative approaches involving stakeholders beyond academia. This idea bridged traditional academic research with real-world applications, providing a practical foundation for TD.

Nowotny et al. (2003) expanded the concept of "Mode-2 knowledge production," focusing on its contextualized and flexible nature. They argued that research must align with societal demands and prioritize accountability to produce socially robust knowledge. Their work advanced TD by integrating stakeholder collaboration and emphasizing the importance of real-world applications. Despite its strengths, the framework offered limited insights into ecological applications and participatory evaluation methods.

Hadorn et al. (2006) emphasized participatory approaches to sustainability, integrating ecological, social, and economic knowledge systems. They highlighted the importance of cross-sectoral collaboration in addressing complex sustainability challenges, positioning TD as a critical framework for such efforts.

Pohl (2010) championed participatory research in TD, advocating for the inclusion of diverse perspectives to address real-world challenges. He emphasized knowledge integration as essential to successful TD research. Pohl advanced the practical application of TD principles by shifting focus toward stakeholder-driven approaches.

Pohl (2011) differentiated TD metrics from traditional academic evaluation, emphasizing societal relevance and participatory methods. He proposed using participatory metrics to assess TD's success in addressing societal needs. This contribution helped refine TD evaluation frameworks but did not fully address long-term impact measurements or provide tools for practical implementation.

Pohl et al. (2021a, b) expanded TD frameworks to include robust stakeholder engagement methods, focusing on integrating diverse knowledge systems in addressing societal challenges. Their emphasis on participatory processes strengthened TD's real-world applications. Although their efforts have been to include the community participants in the TD process, the research teams involved in problem analysis and problem-solving at the community level were only the domain experts. The stakeholders in the problem context were not part of the TD process. The domain experts only consulted the stakeholders to understand the problem and learn to solve it.

Lawrence (2015) explored methodological challenges in applying TD to urban planning and health research. He introduced co-design and co-production approaches to foster collaboration and broaden TD's practical applications. His work addressed practical challenges but did not fully account for the complexities of stakeholder diversity or systemic barriers in global contexts.

Scholz (2002) introduced the embedded case study method, integrating quantitative and qualitative knowledge within TD research. He highlighted the value of combining data-driven and participatory approaches to address complex problems. This methodology enhanced TD's practical applicability but relied heavily on detailed case contexts, limiting its scalability to more extensive systems or global challenges.

5.1.6 Coherence for Effective Integration in TD

Although the efforts of transdisciplinary approaches since their inception were to develop a comprehensive and holistic understanding of complex problems to solve them better, they struggled to provide coherence. Lack of internal consistency and coherence across different levels/sub-systems/dimensions of an evolving reality provided limited applications of TD approaches to specific contexts. TD's bottleneck has lacked coherence in ontology, epistemology, logic, and axiology.

Edgar Morin recognized the lack of coherence and offered this perspective first. Later, Thiery Ramadier, based on his studies in the urban context, subscribed to this view of Morin and applied it to solving problems in the urban context. Building on Nicolescu's work on Hidden Third, Sue, L.T. McGregor argued for effectively integrating contradictions among stakeholders through an axiology of TD values.

Morin (1997) developed the concept of "complex thought," advocating for integrating diverse perspectives to address interconnected societal challenges. His approach underscored the importance of addressing problems contextually, viewing reality as a network of interrelated systems. Morin's work provided a philosophical foundation for TD, emphasizing the need for coherence in synthesizing Knowledge. However, his ideas' abstract and theoretical nature limited their immediate application in practical TD research.

Ramadier (2004) proposed coherence over unity in TD, emphasizing the value of integrating marginal disciplinary insights. His work was particularly relevant in urban studies, offering a framework for tackling complex urban issues through collaborative approaches. Ramadier provided a practical model for TD research by respecting disciplinary differences while promoting integration. However, applying his ideas was narrowly focused on urban contexts, with limited consideration for broader ecological or societal challenges.

McGregor (2018) clarified the ontology, epistemology, logic, and axiology of Nicolescu's Hidden Third. Her effort was to better integrate the contradiction due to plurality in TD, as Nicolescu and other scholars had observed. Contrary to Nicolescu (2010), who said that axiology is unnecessary for TD, McGregor argued that axiology is inherent to every human logic. Her recognition of the need for TD value (axiology) to remove the contradictions aligns with the need for coherence across ontology, epistemology, logic, and axiology.

5.1.7 AIESS-Based ESTAR

Nayak (2019, 2024c) provides an eclectic framework of the transdisciplinary approach with empirical evidence to solve complex problems in different ecosystems, including natural, rural, urban, and industrial ecosystems. Nayak (2024a) discusses the basic design of Ecosystemic Transdisciplinary Action Research (ESTAR), which is elaborated in this paper. ESTAR is based on a nuanced version of systems thinking (AIESS) that provides a coherent basis regarding ontology, epistemology, logic, and axiology of the change we observe in reality (Nayak 2024b). The same framework is used to resolve complex, wicked problems of our evolving reality.

A summary of specific contributions, key concepts, and core assumptions of the different stands of TD within the last fifty years is shown in Table 5.1. The chronology of TD's development exhibits TD's evolution and refinement toward resolving complex problems. The purpose of the latest advancements of TD in terms of ESTAR adds three critical features, viz., ecosystemic perspective to problem diagnosis, action research approach to problem solving, and most importantly, a clear and coherent position of axiology (weakest other-centeredness) for all stakeholders in action and all logical constructions for theory building. ESTAR reemphasizes the systems perspective in its diagnosis and prioritizes ecological balance (environmental justice) in its design for regenerativeness, the basis of hope for all.

Although bringing researchers from multiple disciplines and various stakeholders to problem-solving has become part of TD research, the general understanding of lead researchers in TD research tends to study a problem in isolation from other problems in a system. For instance, a problem, say starvation, is studied from different disciplinary fields, say economics, ethics, and ecology, on the one end, and from various fields of social practices, say business, civil society, and public agencies, on the other end (Hadorn et al. 2006). However, the problem of starvation itself may be linked to other problems in the system, such as land degradation, diseases, poverty, etc. In the above case, the problem of starvation is not studied in relation to the other apparently related problems.

While the involvement of disciplinary experts in TD research has increased, the involvement of various stakeholders from the problem context is not usually included on equal footing in both diagnosing and solving the problem. In transdisciplinary teams, as reported by leading transdisciplinary researchers, members include only the researchers from different disciplines. The stakeholders from the problem context merely serve as information providers to the team of disciplinary experts. Only the researchers are involved in problem framing, problem analysis, and exploring impact (Pohl et al. 2021a, b).

Table 5.1 Chronology and evolution of transdisciplinary (TD) approaches with their key features

Strands of TD	Representative Author(s)	Core Assumption	Key Concepts	Problem Solving Approach
(a) Holistic Knowledge for Problem-Solving	Erich Jantsch (1970, 1972)	Complex problems need integration	Systems theory; Integrative education	Holistic and interdisciplinary
	Jean Piaget (1972)	Knowledge integration improves understanding	Holistic learning; Interdisciplinarity	Comprehensive learning and analysis
(b) Multidisciplinary and Inter-Disciplinary Approaches to TD	Mittelstraß (1992, 1993)	Theory and practice can unify	Unity of science; Knowledge synthesis	Coherent integration
	Rosenfield (1992)	Health issues require integration	Health–social science integration	Collaboration
	Funtowicz and Ravetz (1993)	Uncertainty requires inclusivity	Post-normal science; Extended peers	Participatory decision-making
	Charter of Transdisciplinarity (1994)	Knowledge should transcend disciplines	Ethical integration; Universal principles	Societal and planetary focus
	Kötter and Balsiger (1999)	Methods must adapt continuously	Collaboration; Methodological adaptation	Flexible integrated inquiry
	Morgan Meyer (2007)	Participation improves framing	Framing; Participatory science	Context-based
	Roderick Lawrence (2010)	Sustainability requires pluralism	Epistemological pluralism; Integration	Holistic and integrative
	Jahn et al. (2012)	Reflexive collaboration improves relevance	Reflexivity; Mutual learning	Socially relevant integration
	Helga Nowotny (2024)	Research must be accountable	Mode-2 science; Resilience	Inclusiveness
(c) Philosophical and Theological Interpretations of TD	Giri (2002)	Boundaries limit innovation	Relationality; Transformative dialogue	Knowledge integration
	Kiyashchenko (2017)	Fragmented knowledge needs synthesis	Boundary-crossing; Dialogue	Holistic collaboration

(continued)

Table 5.1 (continued)

Strands of TD	Representative Author(s)	Core Assumption	Key Concepts	Problem Solving Approach
	Osborne (2015)	Disciplines require reflexive change	Dialectical; Transversal thinking	Transformative inquiry
(d) TD Rooted in Systems Thinking	Basarab Nicolescu (2001, 2010)	Contradictions can coexist	Hidden Third; Multidimensionality	Integrative complexity management
	Julie Thomas Klein (2014)	Wicked problems require systems	Systems science; GST	Integration
	Wolfgang Hofkirchner (2017)	Global issues require systemism	Systemism; Holistic paradigms	Systemic coordination
(e) Stakeholder inclusion in TD	Gibbons et al. (1994)	Robust knowledge needs collaboration	Mode-2 production; Stakeholders	Context-driven
	Helga Nowotny et al. (2001)	Diverse perspectives strengthen relevance	Accountability; Real-world application	Ecologically relevant knowledge
	Scholz (2002)	Diverse evidence enriches inquiry	Mixed methods; Embedded cases	Integrated evidence analysis
	Hirsch Hadorn et al. (2006)	Sustainability requires multidimensional integration	Holism; Participation	Collaboration
	Pohl (2010, 2011)	Real problems require inclusivity	Participatory research; Social impact	Iterative social engagement
	Pohl et al. (2021a, b)	Complexity requires engagement	Stakeholder engagement; Integration	Diverse knowledge coordination
	Lawrence (2015)	Practice requires co-creation	Co-design; Co-production	Collaborative
(f) Integration of plurality in TD	Morin (1997)	Reality is interconnected	Complexity theory; Coherence	Adaptive and holistic
	Ramadier (2004)	Diversity strengthens integration	Coherence over unity; Marginal insights	Pluralistic understanding

(continued)

Table 5.1 (continued)

Strands of TD	Representative Author(s)	Core Assumption	Key Concepts	Problem Solving Approach
	McGregor (2018)	Values enable integration	Ontology; Epistemology; Axiology	Coherent perspective integration
(g) AIESS-Based ESTAR	Nayak (2024b, 2024c, 2024d)	Nothingness of any individual entity in a system leading to the weakest other-centered intentionality, and the necessity of coherence in axiology for balance.	Interconnection; Interdependence; Interaction; Intentionality	Ecosystemic Transdisciplinary Action Research (ESTAR) that prioritizes ecological balance

In short, the two significant limitations of the prevailing transdisciplinary research include (a) a social problem, such as starvation, is assumed to be independent of other possibly connected problems, and (b) the stakeholders of the problem context are not directly involved in the problem diagnosis and the synthesis of solutions.

Contrary to the context of research practice under the prevailing transdisciplinary approaches, the ESTAR methodology includes and takes all the stakeholders, viz., researchers or disciplinary experts, and stakeholders from the problem context as equal partners. The details of its implementation are shown through illustrations in the following sections.

5.2 Ecosystemic Transdisciplinary Action Research

The primary research methodologies in different sciences included historical methodology, survey methodology, case methodology, grounded theory methodology, ethnography methodology, and action research methodology. Historical methodology is based on a materialistic perspective, which means material facts could be used to understand the same way people think. On the contrary, survey methodology is based on an ideational viewpoint, which means that by assessing people's ideas and opinions, their material preferences can be understood. The rest of the above-cited four research methodologies, viz., case methodology, grounded theory, ethnography, and action research, have features with different levels of combination of materialistic and ideational perspectives.

Systems science, Transdisciplinary, and Ecosystemic Transdisciplinary are newer additions to the research methodologies. A comparative analysis of different methodologies adopted in various sciences, viz., social, economic, political, and natural sciences, gives a better sense of how ESTAR bears features close to reality and has the potential to resolve complex problems of our reality, ensuring environmental justice and making hope possible.

Each of the above methodologies has its specific methods of data collection and methods of data analysis. Depending on the nature of the data, whether qualitative or quantitative, tools, techniques, and data scaling are adopted to collect data, and analysis is carried out accordingly. Participatory approaches are standard for a better understanding of complex subjective aspects of social research methodologies. In natural sciences, objectivity has been the essence; the research methodology has been largely experimental. Action research in social sciences has been quite similar to experimental research in which the stimulus-response-stimulus approach is used to study a phenomenon.

With growing specialization and reductionism, each discipline gradually evolved into one or the other research methodologies and developed specific data collection and analysis methods. TD developed gradually in this background, from multidisciplinary, inter-sectoral, and interdisciplinary to transdisciplinary, as we observed in the chronology of TD development. By adopting the systems perspective, each TD methodology tried to integrate more fields or dimensions, especially social-ecological, to capture the variations in complex systems of our reality.

ESTAR evolved through various stages, viz., Systemic Action Research (SAR), Systemic Transdisciplinary Action Research (STAR), and Ecosystemic Transdisciplinary Action Research (ESTAR) Methodology. Most importantly, ESTAR incorporates ecological, social, political, and economic dimensions in its methodology.

Methodology includes species methods, tools, and data collection and analysis techniques, as well as the philosophy of science on which a methodology is built. Philosophy of science includes ontology, epistemology, logic, and axiology. The existing methodologies can be distinguished on the criteria of (a) orientation, (b) nature of reality, (c) purpose, (d) nature of data, and (e) nature of data collection.

The orientation of different methodologies varies from being anthropocentric to ecosystemic. Anthropocentric refers to being human-centric, which grew and became prominent from industrialization to the beginning of the twentieth century. Ecosystemic refers to placing the natural environment and ecology at the core of systems analysis. The nature of reality is based on the philosophical position in different methodologies. The position varies from the position of dualism to the position of non-dualism. The purpose of research ranges from problem measurement to problem-solving. Problem measurement primarily involves diagnosis and knowledge-seeking, whereas problem-solving involves synthesis and solution-seeking, which also becomes the basis for authentic Knowledge. The nature of data at one end is frozen in time and space, whereas at the other end, the data is evolving in nature. The nature of data collection varies from etic to emic. Etic refers to a situation where the researcher has not been part of the problem and is an external entity to the problem context; whereas emic refers to a position where the researcher is part of the research problem. Chart 5.1 summarizes graphically the above five criteria of the different methodologies of thinking, doing, and living.

ESTAR's ontology, epistemology, logic, and axiology are rooted in AIESS (all interacting evolving systems science, Nayak 2024a) and are based on the axioms of interconnections, interdependence, interactions, and intentionality. While the first three axioms of interconnections, interdependence, and interactions operate naturally, the fourth axiom, intentionality, is the only degree of freedom of thinking beings to influence change in an ecosystem. Intentionality lies on a spectrum ranging from self-centric to other-centric.

As per AIESS, the changes we observe in reality—whether in ecological, social, political, or economic aspects of an ecosystem—are understood through the interactions between various factors within and among these dimensions. The changes in reality (i.e., ontology) are an outcome of the interactions (the epistemology) of the interconnected and interdependent nature of variables, factors, and dimensions of a system (the logic). The nature of change arising from the above logic and epistemology is a function of the decision-maker's intentionality (i.e., axiology). Intentionality lies on a spectrum ranging from self-centric to other-centric orientations of the actors in the transdisciplinary process.

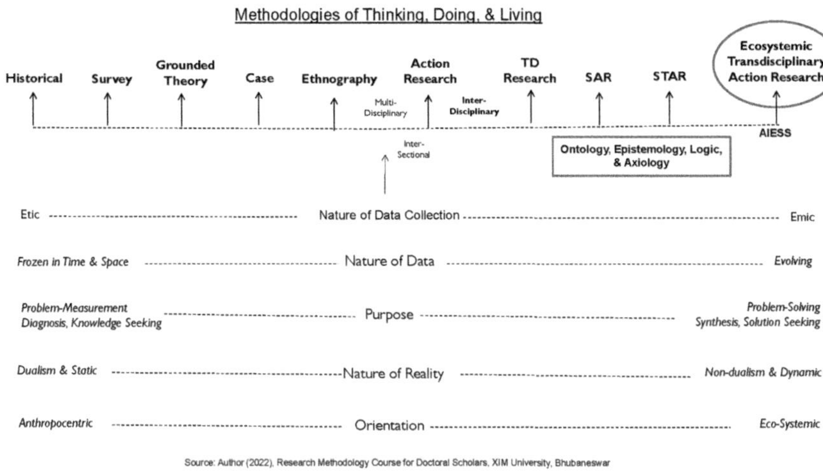

Chart 5.1 Methodologies of thinking, doing and living

In short, Ecosystemic Transdisciplinary Action Research (ESTAR) is a non-dualism based, interacting, and evolving paradigm of thinking, doing, and living in the pursuit of complex problem-solving and knowledge creation. Problem-solving is the primary objective of this methodology, leading to authentic and original knowledge creation. Most importantly, the AIESS-based ESTAR provides an axiology to the philosophy of science that can provide coherence across different fields or dimensions of study. The axiological coherence also provides the logic and clarity to balance human actions at various levels, making ESTAR a practical and effective methodology for solving complex problems and coherent theory development for regenerativeness and hope for all.

5.3 Application of ESTAR in Facilitating Regenerative Agriculture

Let me illustrate here an application of ESTAR in one of the key dimensions of human ecosystems: agriculture. There were two phases of this application.

Phase I—Diagnosis was initiated to recognize the need to restore agriculture from degeneration to regeneration. The five steps of Phase 1, including the identification of critical problems, setting an operational boundary to undertake action research, systemic analysis of relationality among the factors of production in agriculture, dynamic interactions among each set of variables, and delineating degenerative and regenerative cycles among the variables, were carried out. Table 5.2 provides the five key factors of production in agriculture. It also provides two opposite sets of variables that were delineated as degenerating and regenerating sets of variables.

Table 5.2 Key factors of production in agriculture and their sprectrum of variables

		Agriculture		
Inadequate	←	Farm forestry	→	Adequate
Low	←	Farm Diversity	→	High
Unstable	←	Seed-Plant-Animal Genes	→	Stable
Ex-situ & Non-renewable	←	Water	→	In situ & Renewable
Disabling system	←	Soil	→	Enabling System

Diagnosing and identifying the key factors and variables took about six years. It involved deliberations with various stakeholders from the problem context and from outside, viz., farmers, agricultural scientists, sociologists, and policy executives. It involved field visits to farmers across India to study the cases of a few select farmers. Seminars and workshops were also carried out before the action research, and Phase 2 - Synthesis of variables and systemic interventions was undertaken.

Phase II - Synthesis included deliberating and understanding among the stakeholders of the problem context, implementing scientists, lead farmers, and administrators, the weakest link in the agricultural production process. Accordingly, intervention steps were agreed upon, aligning the intentionality of different stakeholders in the action research. Alignment was a gradual process, but it was relatively faster due to the clarity in the logic of interconnectedness and interdependence, and quick empirical evidence through the stimulus-response-stimulus cycles of simultaneous interventions.

In the Phase II interventions, the variables on the right-hand side of the above table, viz., enabling soil system, in-situ and renewable water conservation, stable seed-plant-animal gene, high farm diversity, adequate forest trees, and horticulture trees on the farm were chosen for intervention. The extent of land on which the experiment or action research was carried out was 1 hectare.

The simultaneous interventions on the five critical agricultural production factors, focusing on the soil's weakest life and the weakest life above the ground, initiated regenerative natural cycles in the one-hectare farm. These regenerative cycles resolved the farm's water, carbon, temperature, and crop yield issues within three years of systemic interventions. The experimental farm showed significant improvements in various criteria of in-situ water conservation, soil health (organic carbon, pH, micro-nutrients, carbon sequestration), temperature reduction, check on wind speed, and balanced farm landscape. The cost of operation began to reduce by the third year. Farm income steadily increased to five times the state average within three years. In addition, case studies of farmers across India showed that the net income was as high as INR 80,000 per hectare per month, with an average of about INR 40,000 per hectare per month for the 15 case studies of organic, natural, and ecological farmers in India (Nayak 2024e). This average figure is four times India's national average.

For more details, readers may please watch this video: https://www.youtube.com/watch?v=YXrVJhxz9JE

The improvements in the above criteria or indicators of the experimental farm may be understood from the dynamic, regenerative cycles that were initiated through the simultaneous interventions through the action research. Figure 5.1 exhibits the interconnected regenerative cycles of in-situ water conservation, soil health improvement, diversity, and ecological balance.

The findings from the adoption of ESTAR in agriculture show how the current environmental crisis, a highly complex and globally interconnected phenomenon that is rooted in the rise of temperature and pollution of air, water, and soil systems of the

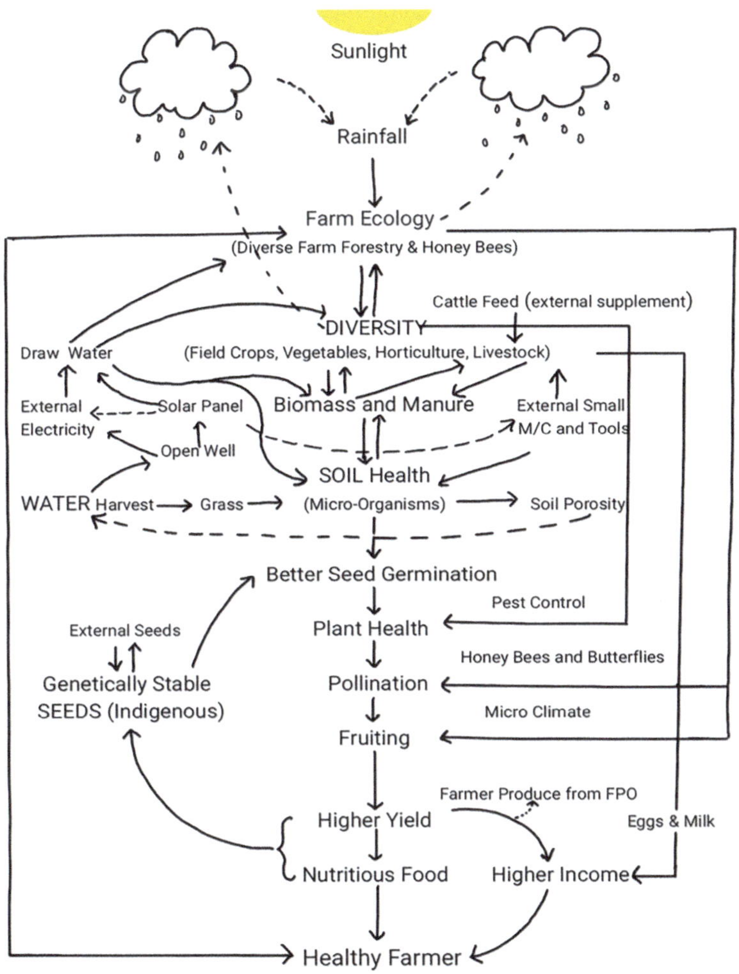

Fig. 5.1 Dynamic interactive and regenerative cycles among key production factors

Anthropocene, could be quickly resolved using ESTAR methodology. This consistent and replicable empirical evidence of regenerative ness and the methodology's coherence of ontology, epistemology, logic, and axiology with reality provides hope for environmental justice and regenerative-ness.

5.4 Relevance of ESTAR in Building Regenerative Ecosystems

ESTAR can also be applied individually to other sub-systems or dimensions of an ecosystems viz., ecology, social relationships, institutions, governance, economic production (primary-secondary, and tertiary sectors), and organizations (Government, non-government, and corporations) for sustainability and regenerativeness. As we saw in agriculture in the above section, a holistic or systemic analysis or diagnosis of the factors affecting agricultural production was the primary step (Phase I). This was followed by identifying the critical variables of degeneration and regeneration, followed by identifying a set of production factors that can facilitate reconstructing or synthesizing regenerative agriculture (Phase II).

ESTAR has been applied to various sub-systems or dimensions over the last fifteen years. The results have been highly encouraging. The broad ecological, social, political, or economic dimensions can be expanded to eight specific dimensions: ecological, relationships, institutions, governance, agriculture, services (including health, education, etc), manufacturing, and organizations. The particular dimensions are further extended to 40 factors (Nayak 2021a, b, 2024c). ESTAR, accordingly, can be applied to diagnose and synthesize regenerativeness in different ecosystems. The lowest unit of governance of a country, such as a Ward, County, Commune, Gram Panchayat, Parishad, Union Parishad, GN in various countries, is the optimal ecosystem for building regenerative ecosystems.

The Government of Odisha in India, in collaboration with the World Bank, has adopted the ESTAR Framework for climate-resilient and market-resilient agriculture at the community level. Six training modules on ecology, social relationships, institutions, governance, agriculture, and community-level organizations have been developed and used to train 100 officials from the departments of the Government to rebuild about 396 PaniPanchayats (PPs) from 15 districts to be regenerative and bring back hope to rural agricultural communities.

Phases and Steps of ESTAR

Using the experiences of systemic action research (SAR) on resolving water and soil issues of a small farm to transform it into a regenerative farming system and preliminary adoption of ESTAR in facilitating Regenerative Ecosystem in a rural-agricultural context for about 20 years, the paper provides a ten-step process from diagnosis of degeneration to resolving the interconnected problems, leading an ecosystem to regenerative-ness. The process of ecosystem diagnosis consists of five steps, and the process of synthesis towards regenerativeness includes another five steps.

Phase I: Diagnosis

1. Identify any critical problem (practical or theoretical) in a system or ecosystem. Any stakeholders may identify the issue in a system or an ecosystem.
2. Set an operational boundary of the ecosystem that broadly entails the phenomenon of the problem. The operational boundary could be a micro-watershed area or a cluster of 2–4 micro-watershed areas, say a Ward, Gram Panchayat (GP), County, etc, that may be the lowest unit of governance in a country.
3. Systematic Diagnosis of the Ecosystem

    3.1 Identify the key actors, factors, and variables of the given phenomenon
    3.2 Identify the relationship networks of these actors, factors, and variables
    3.3 Check for interconnections and interdependence of the key variables.

4. Study the dynamic interactions of the critical variables causing the complex phenomenon in the ecosystem.
5. Delineate regenerative and degenerative cycles or feedback loops created out of the interactions in the ecosystem.

Phase II: Synthesis

6. Deliberate among the stakeholders to identify the weakest link, a person, or any of the stakeholders, factors, or variables in the dynamically interacting ecosystem.
7. Deliberate gently and cautiously to identify the intentionality of different stakeholders involved in resolving the problem. Intentionality refers explicitly to the orientation of a stakeholder towards self-centered, individual gain or other-centered, common good.
8. Deliberate with stakeholders to understand and appreciate the short-term and long-term benefits of both the individual gains and overall benefits in realigning the intentionality of different stakeholders for the common good by strengthening the weakest link of the issue in the ecosystem.
9. Discuss and develop a Strategy for simultaneous interventions on different critical variables to resolve the problem.
10. Start the Action Research regarding the stimulus-response-stimulus approach to facilitate regenerative cycles and resolve the problems. Please see Fig. 5.1

The above steps bear the core features such as interconnectedness, interdependence, and systemic natures of issues and problems in a system, hence the need to recognize the equality of different factors and actors or mutual respect among various stakeholders in an ecosystem. It also bears the features of continuous interactions, dynamicity, and the evolving nature of the system, and, hence, the need for openness of the actors of a system. These features reflect points 1, 3, 4, 6, 10, 13, and 14 of the Charter of Transdisciplinary Research (Nicolescu et al. 1994).

In summary, the Ecosystemic Transdisciplinary Action Research (ESTAR) not only argues for but also provides a clear basis for coherence on two levels viz., (a) coherence across ontology, epistemology, logic, and axiology through its framework of 'all interacting evolving systems science' (AIESS), which currently are not settled in the various transdisciplinary approaches. The above argument and evidence complement and advance the perspectives of a few scholars like Morin (1997), Ramadier (2004), and McGregor (2018), who saw the significance of coherence in settling theoretical and practical contradictions to operationalize TD for effective problem-solving. ESTAR fills this gap in theory and has tested it successfully in practice.

ESTAR offers another level of coherence, viz., (b) coherence across thinking (research), doing (empirical work or practice), and living (alignment of thinking and practice-empirical or theoretical with a reflective living of everyday life). This level of coherence requires balancing human actions by recognizing and adopting the other-centered axiology of ESTAR.

References

Morin E (1997) Complex thought and the need for coherence. Futures 29(6):551–559

Funtowicz SO, Ravetz JR (1993) Science for the post-normal age. Futures 25(7):739–755. https://doi.org/10.1016/0016-3287(93)90022-L

Gibbons M, Limoges C, Nowotny H, Schwartzman S, Scott P, Trow M (1994) The new production of knowledge: the dynamics of science and research in contemporary societies. Sage Publications

Giri AK (2002) The calling of a creative transdisciplinarity. Futures 34:103–115

Hadorn HG, Bradley D, Pohl C, Rist S, Wiesmann U (2006) Implications of transdisciplinarity for sustainability research. Ecol Econ 60(1):119–128. https://doi.org/10.1016/j.ecolecon.2005.12.002

Hofkirchner W (2017) The rationale for complexity thinking and emergentist systemism, ISSN 2309-1606. Philos Educ 1(20)

Jahn T, Bergmann M, Keil F (2012) Transdisciplinarity: between mainstreaming and marginalization. Ecol Econ 79:1–10. https://doi.org/10.1016/j.ecolecon.2012.04.017

Jantsch E (1970) Inter- and transdisciplinary university: a systems approach to education and innovation. High Educ Q 16(3):408–423

Jantsch E (1972) Towards interdisciplinarity and transdisciplinarity in education and innovation. In: Apostel L, Berger G, Briggs A, Michaud G (Eds), Interdisciplinarity: Problems of teaching and research in universities (pp 97–121). University of Nice

Kiyashchenko L (2017) Philosophy of transdisciplinarity: approaches to the definition. Transdiscip J Eng Sci 8:20–35. https://doi.org/10.22545/2017/00084

Klein T (2014) Linking systems science and transdisciplinary research to address wicked problems. Futures 66(1):38–45

Kötter R, Balsiger PW (1999) Interdisciplinarity and transdisciplinarity: a constant challenge to the sciences. Issues Integr Stud 17:87–120

Lawrence RJ (2010) Deciphering interdisciplinary and transdisciplinary contributions. Transdiscip J Eng Sci 1(1):111–116. https://doi.org/10.22545/2010/0003

Lawrence RJ (2015) Advances in transdisciplinarity: epistemologies, methodologies and processes. Futures 65:1–9. https://doi.org/10.1016/j.futures.2014.11.007

Meyer M (2007) Increasing the frame: interdisciplinarity, transdisciplinarity and representativity. Interdiscip Sci Rev 32(3):203–212. https://doi.org/10.1179/030801807X211702

Mittelstraß J (1992) Auf dem Weg zur Transdisziplinarität. GAIA 1:250

Mittelstraß J (1993) Unity and transdisciplinarity. Interdiscip Sci Rev 18(2):153–157. https://doi.org/10.1179/isr.1993.18.2.153

Nayak AKJR (2019) Transition strategies for sustainable community systems: design and systems perspectives. https://doi.org/10.1007/978-3-030-00356-2

Nayak AKJR (2021a) Critical perspectives of public systems management: through the lens of district administration, co-author with Ram Kakani. Routledge. https://doi.org/10.4324/9781003084099

Nayak AKJR (2021b) Simultaneous intervention strategies at local ecosystems for sustainable development goals and peace: design and systems perspectives. In: Oswald Spring Ú, Brauch HG (eds) Decolonising conflicts, security, peace, gender, environment and development in the Anthropocene. The Anthropocene: Politik—Economics—Society—Science, vol 30. Springer. https://doi.org/10.1007/978-3-030-62316-6_23

Nayak, A.K.J.R. (2024a). Regenerative ecosystems in the Anthropocene: a transdisciplinary ecosystemic framework for regenerativeness. Springer https://doi.org/10.1007/978-3-031-53298-6

Nayak AKJR (2024b) Introduction to the AIESS (Ecosystemic) approach to knowing and doing for regenerative ecosystems. In: Nayak AKJR (ed) Regenerative ecosystems in the Anthropocene. The Anthropocene: Politik-Economics—Society—Science. Springer, p 38. https://doi.org/10.1007/978-3-031-53298-6_1

Nayak AKJR (2024c) All interacting evolving systems science: a framework to diagnose degenerative ecosystems and synthesise regenerative ecosystems. In: Nayak AKJR (ed) Regenerative ecosystems in the Anthropocene. The Anthropocene: Politik—Economics—Society—Science. Springer, p 38. https://doi.org/10.1007/978-3-031-53298-6_2

Nayak, A.K.J.R. (2024d). Keynote conference address: from anthropocentric to ecosystemic methodologies: ontology, epistemology, logic and axiology of ecosystemic transdisciplinary action research to resolve the current ecological crisis. In: International conference on environmental justice, Lassalle Institute, Switzerland, August 2024

Nayak, A.K.J.R. (2024e). Documenting cases of successful farmers adopting indigenous, climate resilient and sustainable farming practices in India, NABARD Research Study, p 48

Nicolescu B (2010) Methodology of transdisciplinarity: levels of reality, logic of the included middle and complexity. Transdiscip J Eng Sci 1(1):17–32. https://doi.org/10.1007/978-3-031-70680-6_2

Nicolescu B, Morin E, de Freitas L (1994) Charter of transdisciplinarity. Adopted at the first world congress of transdisciplinarity. Convento da Arrábida, Portugal

Nowotny H, Scott P, Gibbons M (2001) Re-thinking science: knowledge and the public in an age of uncertainty. Polity Press

Nowotny H, Scott P, Gibbon M (2003) Introduction: "mode 2" revisited: the new production of knowledge. Minerva 41(3):179–194

Nowotny H (2024) Re-visioning Mode-2 knowledge: toward societal resilience and accountability. Futures

Osborne P (2015) Problematizing disciplinarity: transdisciplinary problematics. Theory Culture Soc 32(5–6):3–35. https://doi.org/10.1177/0263276415592245

Piaget J (1972) The epistemology of interdisciplinary relationships. In Interdisciplinary education. In: Apostel L, Berger G, Briggs A, Michaud G (eds) Interdisciplinarity—Problems of teaching and research in universities, Center for Educational Research and Innovation (CERI) OECD. Elsevier, pp 127–139

Pohl C (2010) From transdisciplinarity to transdisciplinary research. Transdiscip J Eng Sci 1:65–73. https://doi.org/10.22545/2010/0006

Pohl C (2011) What is progress in transdisciplinary research? Futures 43(6):618–626. https://doi.org/10.1016/j.futures.2011.03.001

Pohl C, Hadorn GH, Lawrence RJ (2021a) Stakeholder engagement in transdisciplinary research: processes, outcomes, and challenges. Sustain Sci 16(5):1451–1463

Pohl C, Thompson Klein J, Hoffmann S, Mitchell C, Fam D (2021b) Conceptualizing transdisciplinary integration as a multidimensional interactive process. Environ Sci Policy 118:18–26. https://doi.org/10.1016/j.envsci.2020.12.005

Ramadier T (2004) Transdisciplinarity and its challenges: the case of urban studies. Futures 36(4):423–439. https://doi.org/10.1016/j.futures.2003.10.009

Rosenfield PL (1992) The potential of transdisciplinary research for sustaining and extending linkages between the health and social sciences. Soc Sci Med 35(11):1343–1357. https://doi.org/10.1016/0277-9536(92)90038-R

Scholz RW (2002) Embedded case study methods: integrating quantitative and qualitative knowledge. Sage Publications

McGregor SLT (2018) Philosophical underpinnings of the transdisciplinary research methodology. Transdisciplinary J Eng Sci 9:182–198

The Charter of Transdisciplinarity (1994) Adopted at the first world congress of transdisciplinarity. Arrábida, Portugal

Amar K. J. R. Nayak Professor of Strategy and Chairperson of the AIESS Research Center at XIM University, Bhubaneswar, India. Drawing on over three decades of field-based action research and policy engagement, he founded the All Interacting Evolving Systems Science (AIESS) paradigm to explain the dynamics of regeneration and degeneration in complex systems. His work on AIESS and its wide application across disciplines, systems, and ecosystems proposes that balance is the fundamental intentionality of nature, making axiological coherence the essential condition for regenerativeness in the dynamically interacting and evolving realities.

Chapter 6
Paths Towards Common Good: Local Knowledge and Transdisciplinarity in Rural Mexico

Diego Orozco Fernández, Carmen Bueno Castellanos, Adriana Carolina Flores Díaz, Madai García González, Ximena Saraí Posadas Vázquez, Sindy Atzhiri Ruiz Cheluja, Alberto Irezabal, and Tania Sánchez Ortiz

Abstract Transdisciplinarity is presented as an approach that questions the way we generate knowledge in Western science, allowing us to recognize the multifaceted worth of practices, feelings, values, and knowledge in a more collective construction that starts at the local level and is a reference that places inclusion (and, therefore, justice) at the center of caring for a territory's viability. Through three experiences with indigenous communities in rural Mexico, this text focuses on the differentiated paths of territorial relations and management as well as the ties created by collective organizations of productive activities and political management in pursuit of their common good. Approaching these experiences creates the possibility to learn about: (a) how knowledge and practices are intertwined, formulated in contexts where such

---

D. O. Fernández (✉) · A. Irezabal · T. S. Ortiz
Social and Solidarity Economy International Research Center (CIIESS), Universidad Iberoamericana, Mexico City, Mexico
e-mail: diego.orozco@ibero.mx

A. Irezabal
e-mail: alberto.irezabal@ibero.mx

T. S. Ortiz
e-mail: tania.sanchez@ibero.mx

C. B. Castellanos
Universidad Iberoamericana, Mexico City, Mexico

A. C. F. Díaz · X. S. P. Vázquez
University Transdisciplinary Center for Sustainability (CENTRUS) Universidad Iberoamericana, Mexico City, Mexico
e-mail: adriana.flores@ibero.mx

M. G. González
Mexico City, Mexico

S. A. R. Cheluja
Innovando la tradición A.C, Oaxaca, Mexico

T. Karcher SJ et al. (eds.), *Making Hope Possible*, Environmental Humanities: Transformation, Governance, Ethics, Law 7,
https://doi.org/10.1007/978-3-032-17719-3_6

knowledge makes sense; (b) the socio-ecosystemic relations present in each experience; (c) the proposals or ties geared towards the common good, and (d) the frictions where the reality demands a reordering.

Keywords Local knowledge · Indigenous peoples and local communities · Common good

6.1 Introduction

Deepening global issues with consequences that affect life in human societies, both currently and in the future, are mainly perceived locally; among them, the loss of biodiversity, changes in climate patterns (IPBES 2019), and a predatory system of production and consumption stand out. Local communities experience the repercussions of these issues in multiple ways which are also critical spaces for reversing such trends, as the territories of indigenous peoples and local communities make up at least a quarter of the planet's surface (Rights and Resources 2023). Initiatives focused on local territories can directly influence the care of biodiversity and ecosystems, simultaneously addressing inequity and poverty (Lennon and Barret 2024) and creating benefits that are transferred from the direct participants and their families to community and regional spheres, also contributing to global dynamics (biophysical and social).

In this text, we analyze the paths taken by three local community organizations toward the common good. We present three cases in which the territory is at the center, incorporating its own features, elements, and dynamics that are a matter of reflection. The geographic diversity, inhabitants, and agreements of ownership and access to resources are some of the elements that bear witness to the practices and values with potential to become transformative innovations.

In all three cases, the relationships with the territory demonstrate that these practices and values have concrete points of reference, which results in decisions aimed—among other things—at recovering the right to one's own way of thinking and feeling about daily affairs and the collective building of futures. Given the multiple ways of generating knowledge, transdisciplinarity (TD) opens up an opportunity to create spaces where diverse actors converge, in which it is common for local experiences to assume a leading role. In this sense, TD receives *the roots and knowledge themselves* as the main source for creating ways of experiencing the territory, which is a meeting point and construction in motion. This analysis is based on the voices of those who build their lives around elements such as hillsides, mountains, forests, fields, and rivers, which are also entities that participate in life and are inter-defined as: (1) the addressed topic, (2) the reference points and validation of methods and results, (3) and the categories of analysis used to reach valid conclusions that are not predetermined but rather set forth based on their relevance (Wanner et al. 2018), thus making space for the group to allow its own new categories of analysis to emerge. This is how we incorporate the TD approach as a way to be mindful of the blind spots of traditional Western science or to keep from falling into the commonplaces of knowledge building not geared toward the common good. Instead, we seek to go beyond disciplines to create knowledge dialogues that help solve social problems.

To understand territorial relations, initiatives with a TD approach emphasizes the ways of reading, regionalizing, and naming the territory, the ways of feeling and perceiving it, and the ways in which knowledge is transmitted, since it is an "epistemic arena." It is within the territorial space that the practices constituting people's livelihoods are carried out, and this is also where the idea of the common good emerges (Alfaro Madrones et al. 2015), which asks us to recognize the connection between individual decisions and collective consequences. The local and traditional knowledge that people and collectives have about the territory is critical for biodiversity conservation, the sustainable management of ecosystems, and the governance of territorial spaces (Malmer et al. 2020), provided that its mobilization and usability can be strengthened in collaborative spaces. Local knowledge and expertise have a holistic nature, which encourages a systemic understanding of the multiple elements and relationships that are present in the territory and make it possible to highlight noteworthy ties, whether for conserving ecosystems and biodiversity (see Tecoanapa case), recovering agroecological practices (see Anaa Witsukj case), or upholding coexistence with nature (see Somos Barro case). It is equally important to recognize the friction in which reality imposes forms of logic and creates tension related to the alternatives that go against the dominant system, but which can also serve as pressure to promote innovation and change.

This chapter offers a view of people's *sentipensar*[1] (Cepeda 2017) related to their territories and of their visions, which have been materializing in productive and territorial defense projects, both linked to epistemic defense, as they allow for reproducing, using, and resignifying knowledge about the structure and dynamics of the places they inhabit. We explore the three experiences from the basis of the following questions: (1) How and based on what experience do we react to the reality that calls us to organize, giving voice to values, practices, agreements, contradictions, and complementarities? (2) What elements "from the outside" can we recognize as materials incorporated into the life-production of these experiences? (3) What are the points of reference associated with the common good and how are they tied to decisions made about the territory (use, access, care, restoration)? (4) How are the elements and dynamics of nature (biodiversity and ecosystems, Mother Earth, etc.) joined to the notion of *common good* in narratives, values, and practices within the territory?

By addressing these questions as a set, it is possible to distinguish a series of paths towards the common good that emerge from the analyzed cases. It is important to recognize that the text reflects the work team's current knowledge and is part of a working roadmap around the concept and the promotion of common good in contexts of the Global South.

---

[1] The nature of *sentipensar* requires a connection from the heart with nature as a whole; understanding the cosmos with all its meanings and senses implies not a pure and simple reasoning—only reasoning—but a reasoning-with with-everything-that-is and with-the-heart), that is to say: to co-reason: corazonar. An alternative translation of *sentipensar* could be *sentient-thinking*

6.2 Our Epistemic Position

This paper presents empirical evidence from specific territories in southern Mexico, which invites us to reflect on the relevance of local initiatives that revalue their ecological links with the territory through productive systems and ecosystem restoration activities. Each experience is briefly presented followed by an account of the relationships with the territory and the proposals for linking with the common good or routes towards it that are still being explored. The document aims to situate local knowledge, perspectives, aspirations, and *sentipensares* as elements in which the common good is present through everyday life.

We maintain a dialogue among different conceptual frameworks. On one hand, the Integral Ecology approach proposed by Pope Francis in his encyclical *Laudato Si'*, proposes a profound paradigm shift in how we understand and engage with the world. It is rooted in five main perspectives: the systemic perspective, the transdisciplinary perspective, the transcendental-transgenerational perspective, the contextual-cultural perspective, and the ethical perspective of care, justice and harmony (CPAL 2022). At its core, Integral Ecology expands our perception of reality by emphasising the interconnectedness of all life systems and the need for holistic thinking. It urges us to approach complex social and environmental challenges through integrated, cross-disciplinary approaches, and invites active participation in the pursuit of both social and environmental justice as interdependent and inseparable goals.

On the other hand, spirituality, understood as embedded in the values, meanings, and knowledge systems that inspire an integral ecology in pursuit of the common good. This project seeks to explore the referents that promote solidarity, cooperation, and the equitable redistribution of benefits, while also inspiring the care and sustainable management of natural resources in each case. Unlike Western or scientific cognitive frameworks, this spirituality permeates everyday life—including work, human–nature relationships, and social fabric—in these communities. It reflects an attitude that centres and defends life, resisting mechanisms of harm, stagnation, or destruction. This worldview does not prioritise power or individual material accumulation, but embraces values such as harmony, balance, meaningful relationships, and a sense of interconnectedness among all things. It is a sensorial form of knowledge and cosmology, embedded not only in texts and memory but also in people's hands, minds, and hearts—serving daily work, passed down through generations, and sustaining the values and relationships that shape cultural identity (Toledo 2005).

The authors are interested in local initiatives that are built in the face of productive, economic, and socio-political difficulties in the territories of indigenous peoples and local communities. We are people who work (i) in academic spaces that approach the analysis of these territories from anthropological, economic, and ecological models, and (ii) we are part of/provide support for community initiatives for improving life. From these spaces, we share the collectives' visions and values, which we have gathered from talks, visits, workshops, and participation in various activities, or by being part of these initiatives. Our group has a clear ethical-political stance, which aims for sustainability to be built from justice (Fig. 6.1).

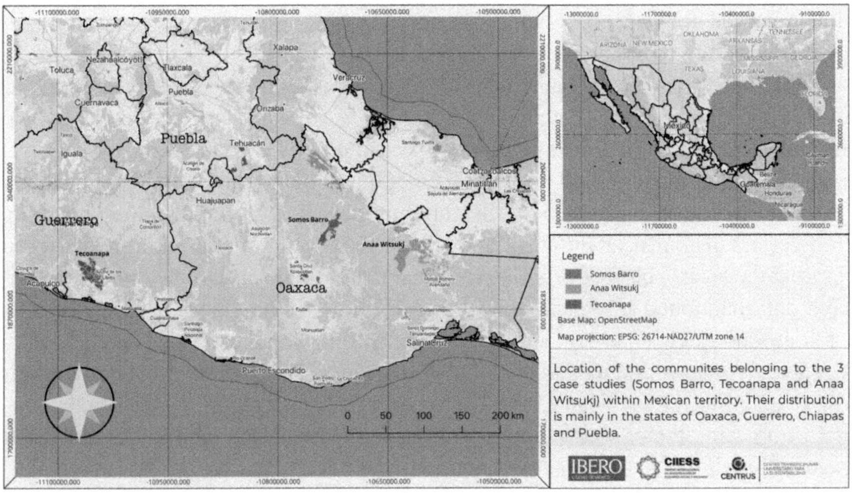

Fig. 6.1 Map showing the locations of the initiatives included in this text

6.3 Experiences of Paths Towards Common Good

6.3.1 Tecoanapa

Since 2005, ten communities in the municipality of Tecoanapa, located in Costa Chica, southeast of Chilpancingo, Guerrero, have experienced significant transformations in their livelihoods due to the contamination of their primary river. This situation was caused by domestic waste, waste from a meat processor, and garbage from the municipal town, located at the top of the basin. When the health of the river and forests declined, the community organized itself to seek solutions designed jointly with a civil society organization. In addition to this problem, the municipality imposed a decision on the communities that water would be brought to the families in the municipal town without the consent of the owners and holders of these forests and resources.

In response, they formed the Tecoanapa Council of Authorities, which is guided by values such as tolerance, truth, and perseverance and gears its actions towards finding solutions for collective well-being. The council's structure includes commissaries representing cooperative land (or ejidos), municipal presidents and secretaries, oversight committees, and former authorities who contribute their experience. The ten communities that make up the council have their own authorities and regulations, reflecting diversity in their internal organizational systems. Decisions are made collaboratively at assemblies, which have an average attendance of between 20 and 30 citizens, and proposals and agreements are documented in minutes signed by the corresponding authorities.

The council members and their families recognize the critical situation that the local ecosystems are facing and understand that their well-being depends directly on the basin's ecological balance. In this context, water—as a human right and vital resource for the ecosystem's functioning—has become the main driver of organization and of connection to the territory.

The creation of a council and the resulting projects represent an effort to guarantee that the communities will remain in the territory. Despite the river's progressive deterioration, the council members continue to seek alternatives for accessing clean water, reducing contamination, and protecting community health. In the process of recovering the river as a living system, the communities of Tecoanapa are implementing biodigesters in homes, an initiative that reflects the population's commitment to recuperating the river. These efforts also reflect a part of their sense of belonging, as they have not abandoned the territory or given up; rather, they are constantly looking for solutions, such as strengthening ties to institutions and external stakeholders to regain their own wellbeing and that of the ecosystem. The community members have also donated hectares of land for the ecosystem's conservation, as they understand that prioritizing the conservation of forests, flora, and fauna supports water catchment within the ecosystem and contributes to the basin's wellbeing. This could mean greater hydraulic availability that benefits nature and its inhabitants.

Another way of relating with the territory is their aspiration to develop a territorial code, considered to be a fundamental step towards the sustainable management of local ecosystems, which will enable them to map the territory and make decisions that benefit both the community and the ecosystem. Its implementation is an expression of making collective agreements for the common good. Also, there is another factor significantly affecting Tecoanapa's prevailing social environment that should be included as part of its environmental deterioration: the insecurity resulting from organized crime that puts its inhabitants' wish to remain in the territory at risk.

The community understands that "water is God's gift to all," that is, they perceive it as a universal resource that should be accessible and equitable. They also recognize education as a fundamental way to promote values of respect for the natural environment and the territory.

To achieve this objective, they have sought to integrate academic actors and institutions that could collaborate in developing solutions to recover the ecosystem's wellbeing. There are ongoing efforts to involve young people in collective concerns. Since its creation, the council has tried to represent all the community's voices under the principle of "there's room for us all," aiming to guarantee the right to life and health and to find solutions for collective demands. However, integrating the indigenous Mixtec population has been a challenge due to language barriers, as the assemblies are conducted in Spanish.

Aiming for long-term sustainability, their initiatives focus on educational practices and political training, to support communities in making informed decisions. This process is developed by seeking solutions through dialogue and a commitment to truth, and by reinforcing the importance of active participation, tolerance, and perseverance in teamwork to achieve the common good. However, its governance also faces challenges, especially in its institutional relations, such as those with the Drinking Water and Sanitation Commission of the State of Guerrero (CAPASEG), the National Water Commission (CONAGUA), and the municipal council, where a lack of transparency and support has been observed.

In Tecoanapa, the conception of the common good is based on the population's active participation in decision-making, an approach that contrasts with the municipal model, which has not addressed their needs. Among the alliances they have formed, their collaboration with Instituto Tecnológico de Estudios Superiores de Occidente (ITESO) stands out—ITESO is a Jesuit university in Guadalajara that proposes a critical sociological intervention focusing on the territory's real and specific needs. In addition, the Council of Authorities is collaborating with another Jesuit university—Universidad Iberoamericana in Mexico City—to regenerate Tecoanapa's ecosystem. Specifically, this is a transdisciplinary effort that has created spaces for integrating local knowledge and experiences with contributions from the academic field to restore the river and conserve the forests, developing actions for the common good.

6.3.2 Anaa Witsukj

Anaa Witsukj (AW) is a collective organization of lime producers who have organized themselves to improve their financial circumstances; they seek a better position in the value chain, strengthen their productive capacities using an agroecological approach, and promote direct trade. After developing strategies to navigate various barriers (Orozco Fernández and Irezabal Vilaclara 2021), in September 2023, AW launched a community packing and trading company in which more than 950 families of producers have participated so far; these families come from 42 communities located in the Bajo Mixe area of Oaxaca, located along the state border with Veracruz.

The Persian lime is a citrus fruit that is consumed around the world, and Mexico is a highly relevant player at a global scale. As of 2019, Mexico is in first place both as a producer and a consumer worldwide, and it is also the second largest exporter (USDA 2024). As a commodity, limes are part of a supply chain that works to satisfy the highest bidder with the aim of maximizing profit. It operates in a market where the limited regulations focus on safety, but there are no policies to control volumes, cycles, or prices to balance the profits of participants, which in turn creates a high degree of volatility based on seasonal cycles but also on hoarding and speculation by some of the chain's actors who have the necessary capacity and resources. There are also networks of intermediaries whose incentive is to buy limes at the lowest possible price to make a profit in the resale process. It might appear to be an uncoordinated chain that reacts to fluctuations in supply and demand, but the actors who benefit from this production and marketing scheme, mainly the large ones, are likely to have underlying coordination efforts, whether they are producers, intermediaries, or marketers.

Lime farming arrived in the communities of Oaxaca about 30 years ago with the creation of a group in northern Veracruz geared at supplying U.S. lime production, which had declined due to weather, pests, and disease (USDA 2014). Unlike other crops, such as corn, limes do not have a history or deep meaning in the culture of Mixe producers and communities, and they are farmed as a cash crop for the market, thus substituting food self-sufficiency and placing them in the value chain's most vulnerable position.

For the purpose of this essay, we will focus on the implications of the underlying practices and values in the management of natural resources for lime production.

Producing high-quality limes requires good soil, water for the trees to retain humidity, and a warm climate. These conditions are found in the AW's territory. In addition, this production occurs mainly on socially owned land; of 347 orchards, 67% are on cooperative farmland (legally designated as ejidos in Mexico) and 17% are communal (the community owns the land and distributes it among its members). Twelve percent is privately owned, and only 4% is rented orchards.[2]

When limes farming arrived in these communities, it brought along a series of agricultural practices seeking to increase productivity by replacing *milpas* (fields using a crop-growing system of corn, beans, and squash) with monoculture lime orchards that employ chemicals to produce higher yields in shorter cycles, which has caused environmental deterioration resulting from soil degradation and biodiversity loss. These types of practices go against those that producers have used for millennia, which seek a balance between ecosystem cycles and sustainable resource management. One of the AW missions is to build capacities and introduce agroecological practices to redirect environmental conditions in a sustainable way.

Since AW is a company inspired by the area's traditional form of community organization and cooperativism, one of its pillars is to care for natural resources and use them in a sustainable manner that guarantees long-term production and does not compromise ecosystems or future generations. In addition, lime production and quality improve significantly when the orchards are well managed, for example by incorporating agroecological nutrition processes to prevent pests and diseases and pruning to induce flowering, in addition to other general care practices, which all benefit the producers.

To achieve this, AW has implemented a multi-stakeholder model in which different actors temporarily contribute resources needed by the project to achieve its objectives. In addition to the founding producers, Universidad Iberoamericana's International Research Center for Social and Solidarity Economy (CIIESS, for its Spanish initials), a private company, social investors, and a group of foundations also participate. The CIIESS has played a key role in this model by coordinating the other actors and even serving as a cultural translator between contexts and ways of living as different as urban and rural or business and community, which has built trust and aligned interests among all the model's participants.

In terms of resource use, AW has been promoting agroecological production practices that producers have been (re)learning and adopting for lime production. This has been done mainly by creating an agroecology school with the collaboration of agronomists, agroecologists, administrative and operational staff from AW, and members of CIIESS and other organizations, such as certifying companies and farmers' associations. In the first group, 35 producers participated for two years. In the second group, which recently began in September 2024, 150 producers are participating, demonstrating considerable interest in strengthening their capacities and adopting sustainable practices.

[2] These data were obtained from a database created by Anaa Witsukj between January and July 2023 with information provided by the producers themselves.

This agroecology school model, with a strong component of farmer-to-farmer exchange, is enriched by the fact that producers share their agricultural practices and those of their ancestors. While agronomists and agroecologists incorporate modern scientific techniques into this knowledge, there is a focus on sustainability that is not merely centered on productivity, and the members of AW, CIIESS, and other organizations contribute key elements from areas such as administration, orchard management, quality management, certification processes, and community enterprises. This is how disciplines are transcended and knowledge dialogues are created to solve social or technological problems. There are several schools of agroecology doing very important work; however, Anaa Witsukj provides a catalyst for profound change since, in addition to promoting sustainable production, it offers producers the option to have control over a packing plant in their territory, which improves their household income and offers a collective benefit that contributes to the common good.

6.3.3 Somos Barro 1050—Cooperativa 1050°

Somos Barro 1050° is a cooperative for the production and commercialization of clay pieces. It was formed in 2019 after a process of over ten years of work, which had a significant focus on participatory design in social and solidarity economy. Its members are mainly female potters who collaborate closely with a civil association called "Innovating Tradition." The knowledge and way of life held by potters has a history spanning at least 4000 years; however, it started to disappear in the 1950s due to migration, the arrival of plastic, and the low appraisal of the trade (Mindling 2023). Today, of the towns in the state of Oaxaca with a potting tradition, 30% are at risk of disappearing, and ways of life, ways of relating to the land, and ways of feeling and thinking about the world (el sentipensar) would be lost with them, as well. In this context, Somos Barro 1050° seeks to dignify the trade, that is, to use cooperative organization to strengthen the community practices that are upheld by generosity and affection as an option for a dignified life. It also seeks to transform and increase the value of the narratives around clay as an economic alternative that provides stability, recognition to potting communities, and hope in life for present and future generations.

Currently, the cooperative is made up of nine potting families. The organization is characterized by the geographic area it covers in southeast Mexico, which includes seven communities located in Oaxaca, Puebla, and Chiapas. Even though all its members do not share the same geographic territory, the cooperative is an example of a collective and inclusive model whose members draw their sense of unity from recognizing themselves as heirs and heiresses of a tradition that has been passed down through generations and, at the same time, is nourished by the multiverse of roots and knowledge that are shared, exchanged and reproduced, in addition to incorporating practices that allow them to earn profits that flow in a redistributive manner for the common good.

For the purposes of this essay, we focus on the first contact with the clay: its site of extraction—both to locate it and to account for handling and conservation practices—which allows us to grasp the great diversity of textures and colors of a resource that comes from the earth. It is precisely the long term persistence of this knowledge that gives the pieces a higher worth than their value in strictly commercial terms because it is heavily steeped in sustainability and tradition. Therefore, scientific and professionalizing knowledge is not present in this reflection. The dialogue between knowledge and interdiscipline appears in the processes of developing and selling the clay pieces. Its appropriation by the cooperative's members has been a lengthy process of accompaniment, innovation, capacity building, and adaptation. However, this feedback between scientific/professionalizing knowledge and the trade makes sense because of the deep connection with the roots of potting life, and this is why we have chosen to explore the extractive phase in greater depth.

Below, we present similarities and differences in the relationship these potters have with Mother Earth in the different territories, the extraction practices, and the values and beliefs incorporated in this first phase, which are the basis for the conception of the common good.

Pottery is a predominantly female trade and most of the cooperative's members are women. Pottery is not usually a solitary trade, as its practices extend to the family setting, taking place in the potter's home-workshop, where sisters, aunts, and grandmothers participate along with parents and children.

To identify where to find the clay, each potter, along with her family, has inherited knowledge that goes back hundreds of generations. Most of the cooperative's members are concentrated in Oaxaca's central valleys, from long-established Zapotec towns such as San Bartolo Coyotepec, San Marcos Tlapazola, and Santa María Atzompa. In San Bartolo, the clay is used in its pure form, and the way it is fired provides its black finish. In the case of San Marcos Tlapazola, red and yellow clays are used. In Atzompa, barro liso (smooth clay) is used with a clay known as barro de golpear (hitting clay or sand), which yields light brown pieces.

The members of the Mixtec village of San Francisco Rio Blanco Tonaltepec are in another region of Oaxaca. Their pieces are the result of mixing the smooth and rough clays they extract from a site called Yusayó—water that comes from under the rocks—and which are characterized by the way they decorate them with a brown color from oak bark. Likewise, in Ixtlán de Juárez, a Zapotec town in Oaxaca's Sierra Norte region, they extract gray clay—also called barro de olla or "pot clay"—which they mix with sand to create light brown pieces.

The bountiful region of the Tehuacán-Cuicatlán Reserve is located in Los Reyes Metzontla, Puebla, where cooperative members use clay extracted from various sites such as the Natuchende hillside, talc from the community's center, and red slip for their pieces, which allows them to achieve different colors and finishes. And finally, in Chiapas, the Tseltal community of Amatenango del Valle is characterized by the white clay that comes from the mixture of sand and quartz, materials extracted from the Cucho Lum mountain, whose name means "to carry clay."

All of these sites, mines, and mountains where clay is extracted are mainly located on communal properties, and some are part of nature reserves. In every case, only community members have extraction rights, and access to the mines is even under surveillance in some towns. Atzompa is the exception since the communal land was privatized as a result of urban expansion, which means the clay must be bought from the owner. Fortunately, peri-urbanization processes similar to the one Atzompa is undergoing near the City of Oaxaca are seemingly unexpected in the other territories since they are located far from cities that are enduring rapid growth.

In general, clay is extracted as a family. In the case of San Bartolo Coyotepec, a custom is kept of only men being allowed to extract it and then distribute it to the community workshops. Collecting the clay is an activity that requires physical strength and rigorous observation of the deposits. The tools used to extract it are similar in every case: picks, shovels, and crowbars; most transport it in sacks. The deposits are located close to the community, and while it does not take longer than an hour to get there, extracting the clay can take a whole day's work. This depends on the amount to be extracted. Except for the family from Mixteca Alta that uses a burro to transport the clay, the others usually rent transportation both to get to the mine and to transport the extracted clay to their homes or workshops.

In this case, the deep-seated quality of coexistence with Mother Earth stands out. There is a dialogue, an understanding, and respect for conserving nature, the milpas, and the animals, which makes it a holistic way of life, essential for supporting families and good living. Throughout the production chain for the clay pieces, the ties to the land are multifaceted. It is a process rich in wisdom that has been incorporated into the dialogue with the material extracted from the mines. The trade of pottery is traditionally manual, that is, throughout the process the hands guide, listen, feel, and think with the clay. Each clay piece that comes from potters' hands is unique from beginning to end—it will not always be the same mountain, the same rain or sun.

In potters' ways of life, patience and spiritual respect are transcendental—only then will the mountain be generous; there is great respect for only taking enough, asking her and her guardians for permission. This practice has given Mother Earth the chance to regenerate the clay that accompanies the families' sustenance. "Before leaving home, we ask God to help and protect us so we can find and take the clay. Since it is early in the morning, we go without eating breakfast. For lunch, we take eggs so we can find the white or gray clay. If we take meat, we do not find the clay, because it's only clay-like soil," shares a member of the community of Amatenango del Valle in Chiapas.

Therefore, being diligent with the mountain, the hillside, or the site is fundamental for the families that make up this cooperative. It is a custom that becomes the essence of a millenia-old trade that has fostered an intimate relationship with the ecosystems that live and feel. Amid a context in which productivity takes precedence over life, its existence and persistence in current times is evidence of long-standing resistance. "Clay connects us to the land and makes us know we are part of her." This testimony, repeated in the narratives of the cooperative members, wisely reminds us and teaches us what is truly important and sacred for life. To strengthen their way of life is to strengthen ways of thinking, social relationships, and relationships with the land, demonstrating the survival of essential traditions for the common good.

6.4 Conclusion

From the integral ecology perspective, the paths experienced in each case have allowed us to compare and contrast relational forms with ecosystems and among community members within the territory. The *sentipensares* are encoded in local knowledge, while also revealing production and consumption habits that emerge with the desire for the common good.

In the analysis of the three cases, we find echoes of the perspectives that underpin the epistemological proposal of integral ecology. The combination of the transdisciplinary and contextual-cultural perspectives allows for the incorporation of diverse knowledge, enriching our epistemological stance and enabling us to guide transformative changes. Therefore, the recognition of local knowledge about essential resources in all three cases highlights the connection between these communities and their resources, directing them toward the common good.

In the Tecoanapa case, water is acknowledged as a vital resource and a human right from a broad perspective: "Water is given by God and belongs to everyone." There is also concern about its depletion and pollution, linked to external impositions, such as regulatory barriers and erratic public policies. Additionally, in this case, water is closely connected to the forest, with a relationship between the forest's water-capturing ability and the springs that feed the rivers.

The AW and Somos Barro cases illustrate how land is perceived as the foundation of life in these territories. Land holds cultural meaning but it also causes external friction due to an extractive model that has dispossessed communities of their lands and resources, leading to conflicts. In this sense, various regulatory regimes have transformed land ownership from communal to private through "legalization" processes.

For the women potters, clay represents an inspiration for life. They assert that "tradition is a thought, not a product," and keep doing ceremonies to seek permission from Mother Earth to extract the clay which is the reason for the potter's life. This women's cooperative demonstrates the love and wisdom they have for their ancient craft, although they now produce goods for the global market. A similar situation exists in the AW case, where producers seek permission from Mother Earth to sow and make offerings to ensure successful harvests, while also producing goods for the global market.

Experiences and initiatives that use TD in specific contexts as a main approach constitute real-world laboratories, which function as spaces for transformative research, generating results for science (or a field of knowledge) and for practice (Wanner et al. 2018). These types of initiatives enable the co-creation of knowledge to mobilize the vast understanding of the territory for decision-making based on local experience (Malmer et al. 2020).

The specificities of the three cases presented here show us the different paths taken towards the common good. These paths give an account of resilient proposals promoted by native communities that are based on a series of social relationships geared at respect for nature, reciprocity, and solidarity (Jiménez et al. 2022). By adding the systemic perspective of integral ecology to the previous two, we were able to regroup the forms of relationship, taking into account diverse configurations that lay the foundations for the paths. The first set of ties are those with nature: Living with and as the land and water is part of building hope, of looking back to ancestral knowledge to preserve or regenerate a resource that has suffered the onslaughts of modernization. These ties are reflected in the ways of naming the hill-sides and approaching the mountains, in how decisions are made about forests to protect the water cycle, or in recovering agroecological practices for food production. The second set of ties, through which the first is formed, is the communality that normalizes cooperation and solidarity in the territory, finding an ally in the values and affective bonds of kinship. These ties become transgenerational transmitters of good living, and they are also cast as bonds between parents-sons/daughters-grandsons/granddaughters, which have different expressions in the three cases. These ties reflect the value placed on family well-being and the importance of the transcendental-transgenerational perspective of integral ecology, which becomes a crucial element that gives meaning to the persistence of resilient practices. Finally, the third set of ties is with social stakeholders outside the territory who play the role of providing support and/or mediation, offering the opportunity to access the knowledge and resources required for the projects to make the desired change. These social stakeholders identify with and agree with the commitment to inclusion and sustainability. The systemic perspective of integral ecology requires incorporating the ties that are embedded in these territories and that allow them to establish positive connections with social actors outside the territory.

As part of this commitment to the common good, we also find points of friction and liminal situations where the reality must be reordered. This becomes a driving force for innovating and raising awareness. In ecological terms, AW and Tecoanapa are experiencing a critical phase due to the deterioration of land and water resulting from practices that have been part of "modernizing" the territory. In the case of AW's territory, the use of chemical products to increase lime orchard productivity stands out; and in Tecoanapa, the poor water quality is a consequence of using the basin as a drainage system for dumping domestic and agro-industrial waste. In their purposes and values, both cases include recognizing traditional practices for land and water regeneration and implementing ways of collectively governing the resources that are essential for the well-being of the territory's residents and the ecosystem they inhabit. Further than that, in Tecoanapa there is a territorial perspective that goes beyond the contaminated bodies of water, which is why the development of a territorial code based on the awareness of ties with the forest, hills, and water has been proposed. AW also has a territorial perspective that goes beyond agroecological practices and seeks to strengthen the autonomy of communities to enable them to live according to their own ways of organization and fulfillment.

To achieve the desired changes, AW and Tecoanapa have sought out support from social stakeholders who provide scientific knowledge and other resources in search of an agroecological transition. These ties, under a model of multi-stakeholders or consulting services on environmental issues, have found fertile ground in the ancestral practices that are still alive in these communities' memory and values of coexistence.

The role of universities as advisors or mediators to guide these projects towards the path of the common good stands out.

A contrasting case is Somos Barro, whose strength is in the survival of over 5,000 years of ancestral roots, made possible by keeping the holistic and spiritual ties with their territories alive to extract the resource that is most important for the process and that gives an integral value to the products that are allocated to cosmopolitan trade channels. Clay—as a vital resource—and its forms of extraction are an essential part of life. Through it, one can reflect on the high sense of communality, integration, and identity that the families in charge of the entire process carry with great pride and that is the source of inspiration guiding all the actions of this cooperative.

These ties and frictions are part of the paths toward common good that emerge locally in regions where communities have had to face the diverse consequences of globalizing contexts collectively. They offer clues about the options, practices, and sentipensares that are within their reach to manage natural resources from a harmonious and regenerative perspective as well as the transdisciplinary commitment to building knowledge that, placed at the service of the territory, has the potential to solve complex problems at the margins.

Although this project is in its initial phase, the progress achieved thus far allows us to continue the collective exploration of the intersections between the ties and frictions in diverse paths toward the common good, which offers us the chance to improve this knowledge and, therefore, local practices aimed at transforming realities. This is how the epistemological change that integral ecology proposes allowed us to recognize and incorporate the explicit position of the ways of feeling and thinking (sentipensar) about the local territory, putting locally led experiences and action at the center and responding to interests and contexts that make sense to the participants, on their own terms.

References

Alfaro Madrones. JI, Fernández Hernández CJ, González García MJ (2015) La transdisciplinariedad, una herramienta para apuntar al Buen Vivir. Polis Rev Latinoam 40. http://journals.openedition.org/polis/10896

Cepeda J (2017) The problem of being in Latin America: approaching the Latin American ontological sentipensar. J World Philos 2(1):1. https://doi.org/10.2979/jourworlphil.2.1.02

CPAL (2022) Guidance framework for studying and working with integral ecology, Casa Leiria. https://jesuitas.lat/archivo/biblioteca/biblioteca-cpal/archivo-documental/guidance-framework-for-studying-and-working-with-integral-ecology-integral-ecology-peer-group-cpal-social-centers-network

IPBES (2019) Global assessment report on biodiversity and ecosystem services of the intergovernmental science-policy platform on biodiversity and ecosystem services. https://doi.org/10.5281/zenodo.3831673

Jiménez A, Delgado D, Merino R, Argumedo A (2022) A Decolonial approach to innovation? Building paths towards Buen Vivir. J Dev Stud 58(9):1633–1650. https://doi.org/10.1080/00220388.2022.2043281

Lennon J, Barrett B (2024) Conserving an underappreciated heritage resource: the rural landscape. Parks Stewardship Forum 40(3):562–576. https://doi.org/10.5070/P5.35442

Malmer P, Masterson V, Austin B, Tengö M (2020) Mobilisation of indigenous and local knowledge as a source of useable evidence for conservation partnerships. In: Vickery JA, Ockendon N, Pettorelli N, Brotherton PNM, Sutherland WJ, Davies ZG (eds) Conservation research, policy and practice, 1st edn. Cambridge University Press, pp 82–113. https://doi.org/10.1017/9781108638210.006
Mindling, E. (2023). Barro y fuego. El arte de la alfarería en Oaxaca 4. Innovando la tradición
Orozco Fernández D, Irezabal Vilaclara A (2021) Evitando el fracaso ante la crisis: los casos de dos organizaciones indígenas y campesinas de la economía social y solidaria en Oaxaca y Chiapas. Sobre México Temas de Economia 1(2):5–44. https://doi.org/10.48102/rsm.vi.66
Rights and Resources (2023) Who owns the world's lands? Global state of indigenous afro-descendant, and local community land rights recognition from 2015–2020, 2nd edn. https://doi.org/10.53892/MHZN6595
Toledo VM (2005) Los curadores de la madre tierra (un manual etnoecológico para los pueblos tzeltales del norte de Chiapas). UNAM – Misión Bachajón – CEDIAC
USDA (2014) Fruit and tree nuts outlook: economic insight. https://ers.usda.gov/sites/default/files/_laserfiche/outlooks/37059/49130_fts-357.pdf?v=43812
USDA (2024) Citrus: world markets and trade. https://fas.usda.gov/sites/default/files/2024-07/citrus.pdf
Wanner M, Hilger A, Westerkowski J, Rose M, Stelzer F, Schäpke N (2018) Towards a cyclical concept of Realworld laboratories: a transdisciplinary research practice for sustainability transitions. disP The Plan Rev 54(2):94–114. https://doi.org/10.1080/02513625.2018.1487651

Diego Orozco Fernández is part of the International Center for Research on the Social and Solidarity Economy (CIIESS) at the Ibero-American University, Mexico City. He holds a degree in International Trade and a Master's degree in Governance, Development, and Public Policy. His work focuses on the Social and Solidarity Economy in connection with development from a territorial perspective, in collaboration with projects led by social and solidarity economy organizations, with a particular focus on peasant and indigenous communities working in the agro-industrial system.

Carmen Bueno Castellanos holds a PHD in Anthropology and is Level III member of Mexico's National System of Researchers and a member of the Mexican Academy of Sciences. Her work focuses on the anthropology of organizations, labor, and globalization. She is the author of influential publications on productive systems, innovation, and global transformations.

Adriana Carolina Flores Díaz holds a degree in Biology and Ecology, and designs research and action projects on rivers and riverbanks, water monitoring, and local water governance systems. She is a co-founder of Global Water Watch Mexico, dedicated to strengthening citizen capacity for watershed monitoring since 2004. She participates at global platforms such as IPBES since 2018, and leads Riberear, a network dedicated to reconnecting communities and rivers. Her publications address topics related to water monitoring, riverbanks, and nitrogen deposition in rivers. She is based at Universidad Iberoamericana, and recognized as National Researcher (SNII-1) in Mexico.

Madai García González holds a master's degree in social anthropology, specializing in socio-environmental movements, the political participation of Indigenous women, and environmental justice. Her work centers on participatory research, which has led her to collaborate with communities defending their territories against large-scale mining and wind energy projects. She has also participated in multidisciplinary collaborative projects with cooperatives and farmers from a Social and Solidarity Economy perspective.

Ximena Saraí Posadas Vázquez holds a Bachelor's degree in Environmental Sustainability. She has collaborated on research, public policy, youth environmental activism movements, ESG reports and territorial diagnosis projects, consolidating a transdisciplinary profile. As an Executive Professional at the Instituto Nacional de Ecología y Cambio Climático (INECC), she contributes to the formulation of public policy instruments—including Mexico's first National Adaptation Policy—and manages strategic coordination for high-level decision-making, strengthening the link between climate change adaptation, environmental governance, and community-level sustainability initiatives.

Sindy Atzhiri Ruiz Cheluja is Strategic Director of Innovando la Tradición A.C. and Operations Coordinator of the Somos Barro 1050 Cooperative, Mexico. She holds a bachelor's degree in Public Accounting and is a graduate-level candidate in the Master's Program in Creation and Development of the Social and Solidarity Economy. Her work focuses on community economies and the ways they inspire and strengthen socio-economic models in urban contexts. Over the past nine years, she has contributed to the development and economic sustainability of the Somos Barro 1050 Cooperative, while supporting the design of a twenty-year collective vision through the creation of a long-term life plan.

Alberto Irezabal Director of the International Center for Research on the Social and Solidarity Economy at the Ibero-American University, Mexico City. Trained as an industrial engineer and holding a PhD in Advanced Management of Organizations and Social Economy, his work focuses on social and solidarity economy, community-based development, and sustainable value chains. Drawing on more than a decade of collaboration with indigenous communities in southern Mexico, he explores pathways for fostering inclusive economic development.

Tania Sánchez Ortiz is a biologist that has collaborated on various national and international ecological and socially engaged research projects, consolidating an interdisciplinary profile that integrates scientific knowledge with sustainable territorial practices. As a researcher at the International Research Centre for Social and Solidarity Economy (CIIESS) of Universidad Iberoamericana, Mexico City, she contributes an ecological perspective to the transformative processes driven by social and solidarity economy organizations, strengthening the link between environmental sustainability, common good, and alternative economic practices.

Chapter 7
Networks that Build Hope in Wounded Territories

Diana Patricia Giraldo, Alberto Irezabal, María del Mar Magallón, Álvaro Idárraga, and Erminsu David

Abstract The Comparte Network, born within the framework of the Conference of Jesuit Provincials of Latin America, is an international network that aims to foster hope amid multiple crises in vulnerable, rural territories across Latin America. It is grounded on the premise that reducing inequality, poverty, and injustice requires subordinating the economy to the pursuit of *Buen vivir* [Good living]—which includes improving social, political, ecological, and cultural dimensions within communities and their habitats. Through a transdisciplinary and multi-actor approach, the network has collaboratively developed its own methodology proposal focusing on territorial development. This article draws on the analysis of the exchange of experiences between two of its members and highlights the contribution of networks to the nurturing of hope in vulnerable and complex contexts. The first part of this article outlines the theoretical foundation of subordinating economic development to the holistic vision of *Buen vivir*. The second section describes the transdisciplinary methodologies developed to support collaborative knowledge generation and action. Next, the third section examines a paradigmatic case from the Comparte

D. P. Giraldo (✉)
Institute of Social Sciences - Development Studies, Lisbon University, Lisbon, Portugal
e-mail: diana.giraldo@edu.ulisboa.pt

A. Irezabal
Social and Solidarity Economy International Research Center (CIIESS), Universidad Iberoamericana, Mexico City, Mexico
e-mail: alberto.irezabal@ibero.mx

M. M. Magallón
Entrepreneurial and Territorial Competitiveness, Innovation and Sustainability, Deusto Business School, Bilbao, Spain
e-mail: Mariadelmar.magallon@deusto.es

Á. Idárraga
Executive Secretary of the Comparte Network, Buga, Colombia
e-mail: sria.ejecutiva.comparte@cpalsj.org

E. David
Instituto Mayor Campesino - IMCA, Coordinator of the Comparte Network, Buga, Colombia
e-mail: direccion@imca.org.co

T. Karcher SJ et al. (eds.), *Making Hope Possible*, Environmental Humanities: Transformation, Governance, Ethics, Law 7,
https://doi.org/10.1007/978-3-032-17719-3_7

Network: the exchange between rural communities involved in the organic coffee value chain in Mexico and Colombia. Finally, the fourth reflects on how collective experiences can translate theological hope into transformative practices that generate lasting change and foster virtuous cycles in wounded territories.

Keywords Buen vivir and alternative economic models · Transdisciplinary methods · Territorial development and integral ecology · Multi-actor collaboration and social innovation

7.1 Introduction

In recent decades, in Latin America and in many regions of the Global South, the situation of violence and social and environmental conflict has deteriorated, especially in rural territories. The impact on the livelihoods and ways of life of local indigenous and rural communities, in particular, has been severe in terms of the tearing of the social fabric, loss of cohesion, and degradation of the natural ecosystems that sustain life (Fuentes 2023).

In response to this unprecedented crisis, many communities and groups of people have organized to meet their different needs through the promotion of alternative economic initiatives (Esteves 2021). According to Hillebrand and Zademach (2013), it is increasingly common to use the term “alternatives” to refer to economic practices or models that diverge, to a greater or lesser extent, from capitalist economic models. In many cases these alternative economic initiatives are often framed within what is known as the social and solidarity economy. According to UN General Assembly Resolution 77/281,[1] this economic model is underpinned by the principles of voluntary cooperation and mutual assistance, democratic or participatory governance, autonomy and independence, and the primacy of individuals and social goals over capital in the distribution and use of surpluses or profits. Various organizational forms are adopted by this economic model, but all of them put into practice a set of values consistent with caring for people and the planet, equality and equity, interdependence, self-governance, transparency and accountability, and the attainment of decent work and dignified livelihoods.

The goal of most of these initiatives is to foster *Buen vivir*, which is understood as a local conception of a lifestyle and a way of understanding the world based on the worldview of the ancestral Andean people (Altmann 2017). This lifestyle seeks to put at center stage: life, communitarianism, and harmonious relationships with nature, territory, and other beings (Irezabal, A. 2023). According to Rodríguez A. (2016), *Buen vivir* represents a paradigmatic proposal and an opportunity to deconstruct the conventional paradigm of development and substitute it for pluralistic proposals that converge with other schools of thought, such as deep ecology, the

[1] General Assembly Resolution 77/281 (18 April 23)

social and solidarity economy, degrowth, and ecofeminism, which seek alternatives to the prevailing development model based on economic growth.

Most of these initiatives are developed in contexts of great precariousness in terms of capacities, access to markets, or financing. The setting up of support networks contributes to communitarian resilience and sustainable growth through the exchange of experiences, access to resources, or the promotion of innovation (Borda-Rodriguez et al. 2016; Kim et al. 2020; Parida and Wincent 2019).

The Comparte Network was set up in 2010 as a space for coordination to strengthen and scale up[2] alternative economic initiatives in Latin America. Through its territorial perspective with a multi-stakeholder approach, Comparte has been supporting the transformation of territories through a transdisciplinary approach to learning and action based on a methodology proposal designed collectively by the Network members. Strategically, the Network promotes inter-cooperation to build capabilities and foster innovation, while strengthening integral ecology through the promotion and development of agroecology.

The Network's contribution to building hope is reflected in the reality of the economic and production experiences it supports. Given their diversity and large number, in this article we present the specific experience of inter-cooperation between two groups: the Mexican cooperative *Capeltic* and the Colombian social enterprise *Garittea*. Both are designed to serve peasant and indigenous families and their communities based on a perspective of reconciliation with nature and the promotion of the reconstruction of the social fabric from a local worldview, all while simultaneously participating in different markets. Likewise, these organizations seek to contribute to the food security and sovereignty of families and to contribute to local and global food systems.

7.2 The Comparte Network in Context

The Comparte Network comprises 18 Social Centers[3] (SCs) within the framework of the social sector of the Conference of Jesuit Provincials of Latin America (CPAL). The SCs work with around 500 economic initiatives (EIs) involving approximately 56,000 families in 12 Latin American countries. The main objective of the Network is to achieve conditions of *Buen vivir* for people and communities in situations of exclusion and poverty by strengthening sustainable economic alternatives that can generate a territorial impact in Latin America.

---

[2] For the Comparte Network, "scale up" is not only about expanding or covering the width of a territory, but also integrating the rating of the initiatives, the generation of value, and their capacity to improve living conditions in a territory, i.e. scaling up also has to do with systemic height and spiritual depth.

[3] 18 social centers in 12 Latin American countries and 1 in Spain

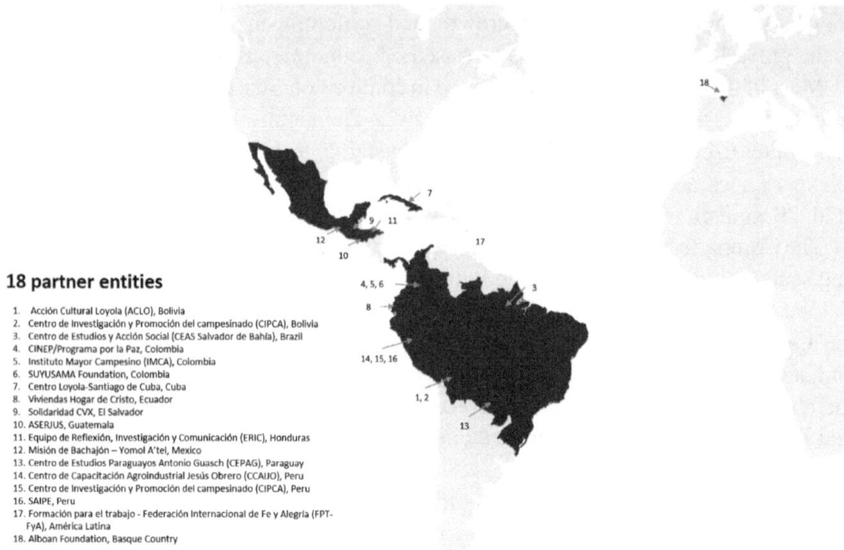

Fig. 7.1 Map showing the presence of the Comparte Network. Red Comparte (2023)

In this respect, Comparte contributes to the recognition of the existence of multiple economic models (popular, solidarity, community, etc.) in the Latin American context. These models nevertheless remain invisible despite the tenacity and resilience of the people and communities who undertake them under conditions of exclusion. By representing a source of real hope for people in situations of exclusion and poverty, this diversity of alternatives, which shares features with socio-environmental justice, challenges the dominant economic system that has brought society to this civilizational crisis (Fuentes 2023) (Fig. 7.1).

Comparte is present in a rich and varied universe of initiatives and projects: local economies, production initiatives, entrepreneurship hotbeds, social enterprises, financing funds, and mixed public policy programmes, among others. These initiatives are promoted by popular, indigenous, afro-descendant and peasant communities, and in diverse regions such as the Altiplano, the Sierra, and the coast. Experience in the midst of such great diversity has revealed common perspectives and challenges in generally paradoxical situations: communities lacking well-being, which are located in territories of abundance that progressively lose that asset, which, in turn, further impoverishes those who are already dispossessed (Aguilar and Idárraga 2017).

Therefore, there are wide varieties of economic initiatives and social centers supported, with diverse experience in the territories: the more than 60-year-old Instituto Mayor Campesino in Colombia, for example, or the more recent organization Tinidia, born from the deseased Nitlapan, which was parte of the Universidad Centroamericana in Nicaragua. What unites them all is their link to the territory, expressed as links with their history, their identity, and the biocultural heritage that defines each of them.

Since its inception, the Comparte Network has built up alternative features that characterize these economic alternatives. These features represent the end goal toward which support for the Network's initiatives focuses, with the aim of contributing to the achievement of *Buen vivir* for people and their communities (Fig. 7.2).

The work of the Comparte Network is characterized by a learning/action approach with a territorial perspective, using a transdisciplinary framework. In other words, it is based on the practical experience of the community in transforming reality (Merçon 2022). It combines a systemic approach to learning, research, knowledge, and reflective action. This approach includes the concept of integral ecology, since it understands that economic, social, and cultural activities are part of a broader natural system: the territory. This unique and complex physical and symbolic space includes different interconnected and interdependent systems (natural, social, political, cultural, economic, etc.) that constantly interact with each other and evolve (Deane-Drummond and Deneulin 2021). Under these terms of reference, the Network has been able to generate its own methodology proposal, integrating methodologies, tools, and approaches which that the coordinated work carried out at various levels: the Network, social centers, and producer organizations and communities where its impact is felt.

7.2.1 The Territory as a Space for Transformation

When defining territories/regions/zones, it is necessary to start from the understanding of micro/local spheres and their various components (social, political, economic, environmental, and cultural), in order to find complementarities. These components at first seem independent, but are gradually incorporated into a broader vision. In this respect, a region/territory is a life space. It is made up of environmental, political, economic, cultural, and social dynamics, in which are combined actors and processes seeking to build conditions of sustainability from their potentialities and challenges.

The territory is the starting point for the intervention and methodology proposal of the Comparte Network. Its proposal for the *Lectura Estratégica del Territorio* (LET, Spanish for Strategic Reading of the Territory) represents the beginning of the process to construct economic and production alternatives for a region (Aguilar and Idárraga 2017).

The LET goes beyond the type of understanding of reality that ends up making a "shopping list" (a pejorative term), in which a large number of things to do are identified without specifying priorities, understanding relative importance or logic, or ordering the intervention. The strategic aspect of LET has to do with how we understand each territory at a given time: by identifying its potentials (heritage,

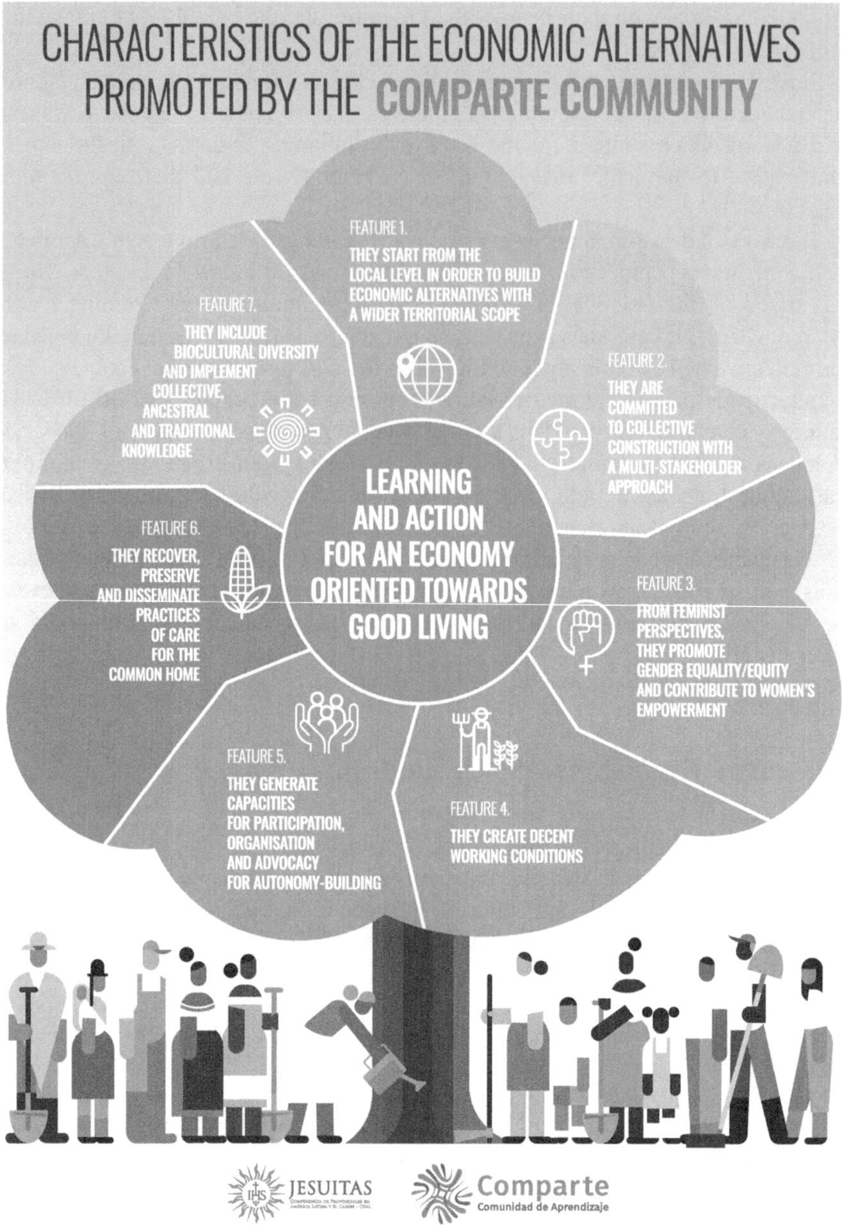

Fig. 7.2 Tree of the features of economic alternatives. Red Comparte (2022)

talents), its difficulties, and its challenges, and then prioritizing plans, programmes, and projects (Aguilar and Idárraga 2017).

In addition to the territorial perspective, the Comparte Network also follows the ideas of Integral Ecology, an approach championed by Pope Francis, which highlights the interconnectedness and interdependence of all living creatures, and which is understood as an alternative to what Pope Francis calls the "modern technocratic paradigm" (Deane-Drummond 2023). In the encyclical Laudato Sí, Pope Francis declares:

> *These ancient stories, full of symbolism, bear witness to a conviction which we today share, that everything is interconnected, and that genuine care for our own lives and our relationships with nature are inseparable from fraternity, justice and faithfulness to others* (Francisco, 2015. n.52).

By considering historical identity in the territories of action, as well as a perspective of interconnection between all living beings, we must necessarily include spirituality in the approach of social centers and economic initiatives. This spirituality fosters interconnection and trust, and, therefore, more horizontal relations within the territory, enabling the creation and development of economic alternatives that emerge from the base.

The concept of Integral Ecology proposes a paradigm shift, an epistemological break with what is denounced in analyses of the ecological crisis. The recasting of an embodied and committed ecological spirituality is called for. This perspective is not purely theoretical or intra-ecclesial. On the contrary, it directly affects the structures that organize society and its political, economic, cultural, social, and environmental dynamics (RCS-CPAL 2022).

7.2.2 *Transdisciplinary Methods of a Learning Community*

When considering the territorial perspective and the Integral Ecology approach, the methodology proposal of Comparte is clearly framed by transdisciplinarity. The construction of economic alternatives that make *Buen vivir* possible is a challenge that can only be addressed in this way. As Nicolescu (1996) points out, we are faced with "the imperative need to propose, live, learn, and teach complex thought that reweaves the disciplines as a possibility of humanity in completeness." This need to transcend disciplines and address complexity arises from the Network's initial proposals, for example, in the collective construction of a methodological roadmap that incorporates diverse voices and knowledge, taking into account the practices and experiences of different territories, each with its owns levels and rhythms. The methodological roadmap does not aim to unify criteria, but rather learn and gather knowledge that people and organizations have generated on the ground and make it available to the entire Network.

Since the Network is a learning community born out of real and concrete economic initiatives and the social centers that support them, transdisciplinarity emerges naturally and is essential for addressing the complexity present in reality. In other words, based on Comparte's experience, when reality is the starting and end point that guides the generation of knowledge and the proposal, it is essential to approach it in a transdisciplinary manner. To this end, the Network has opted to develop a multi-stakeholder approach based on the diversity of experiences and knowledge. This has allowed it to acquire various capacities and perspectives when addressing both the form of and support for implementation of its methodology proposal.

Transdisciplinarity is specified, therefore, in the methodologies, tools and approaches that guide the coordination work carried out by the Comparte Network at different levels (producer organizations, social centers, and the Network).

The various components comprising the network's methodology proposal are set out in the table (Table 7.1).

7.3 A Specific Experience of Inter-Cooperation

One of the key strategies employed by the Network to open horizons, promote innovation, and scale up economic initiatives is inter-cooperation among its members. This strategy is rooted in one of the principles of the social and solidarity economy and shapes Comparte's way of working (Magallón et al. 2022).

As mentioned in Aguilar and Idárraga (2017):

> … The challenge is to move from successful experiences of micro coverage to proposals of subregional and regional impact that contribute to the construction of sustainable economic and production alternatives, as a fundamental condition for the construction of social alternatives in which a greater number of people and families can participate and benefit.

This article presents one from among the diversity of inter-cooperation experiences promoted by the Network. It deals with the inter-cooperation experience between two organizations: Capeltic (Mexico) and Garittea (Colombia). It shows the potential offered by networks as spaces for inspiration and hope through the generation of knowledge and the exchange of experiences among their members (rural and urban communities). It also reveals the approach of the various stakeholders (universities, private companies, consumers, and the public sector), with supporting actors that accompany the process of building new systemic realities in the territories.

The first experience, Capeltic, is a social enterprise comprising five coffee shops that are part of the Yomol A'tel social and solidarity economy cooperative group in the state of Chiapas, southern Mexico. Yomol A'tel is a coordinated, integrated group of social and solidarity enterprises made up of Tzeltal indigenous communities.[4] Workers and collaborators work on the production, transformation, and marketing of organic coffee, honey and its derivatives, microfinance, and territorial development

[4] In the Tzeltal context, communities are small villages or towns made up of several family clans. Each community has its own structure of government, festivals, and social organization.

Table 7.1 Methodology proposal

Methodological proposal Comparte Network	
Methodology/tool	Description
Methodology for capacity building[1]	It includes a map (dimensions and categories) of individual, collective and systemic capacities necessary for the generation of alternative economies in a territory.
Methodological route for the construction of economic-productive alternatives with local and regional impact	Aimed at the identification and planning of economic-productive alternatives with the potential to generate economic and integral transformations from specific territories. It contains a component of strategic Reading of the territory (LET), a strategic framework and monitoring, evaluation, baseline and systematization of the prioritized processes (Aguilar, J.A. & Idárraga, A., 2017).
Social business plan (SNP):	It is an analysis, planning and management tool at the service of the development of economic-productive initiative in all phases of the value chain. The PNS explains how territorial potential is managed in terms of the complete economic cycle by integrating key allies.
Participatory guarantee system (GSP):	It offers a methodology to generate articulated work between agroecological producers and consumers and allows "certifying" what is given is not properly a certification but an action of trust to agroecological production in the plots.
Socio-environmental audit system (SASA):	It measures the degree of implementation of the 7 features of the economic initiatives accompanied by the social centers of the network. It consists of about 76 indicators and is voluntary and annual to complete.
Gender organizational self-diagnosis tool:	Its objective is to facilitate processes of review, reflection and analysis from the gender equity approach in an organization, especially oriented to organizations in the alternative economic and productive field.
Solidarity economic circuits (CES) methodology	It is aimed at creating spaces for the articulation of the practices and actors of the alternative and solidarity economy in a specific region. The CES aspire to cover aspects such as healthy production, solidarity finance, fair trade, responsible consumption, post-consumption, community tourism, ancestral health to satisfy fundamental needs of the human being and build a society of culture of Buen vivir.

(continued)

Table 7.1 (continued)

Methodological proposal Comparte Network	
Methodology/tool	Description
Multi-stakeholder approach to work	The consolidation of sustainable and scalable economic-productive alternatives will only be possible as a result of alliance building processes and the sum of joint commitments and efforts. In this way, the various actors[2] contribute from their field of specialization and knowledge to the sustainability of economic-productive initiatives.
Full business cycle approach (CEC)	The idea is for economic initiatives to advance in the consolidation of value chains, from production to the social reinvestment of surpluses[3]. It requires the definition of strategies that lead to obtaining a high quality of products through processes that allow control and autonomy within the value chain, emphasizing that success will be measured by the quality of the social reinvestment that is achieved.

projects. Its origins date back to the work begun by the Mission of Bachajón of the Society of Jesus in the northern region of the state of Chiapas in 1958.

The second prominent experience is the Garittea social enterprise, which began in 2015 in the Valle del Cauca region, Colombia. Garittea, comprising two associations of coffee growers from Colombia[5] and an NGO,[6] with the support of the Javeriana University Cali, is a social enterprise in charge of marketing agroecological coffee in cups. In other words, the coffee is sold to the end consumer, which has allowed a range of skills to be developed throughout the coffee production cycle.

This participation in the coffee value chain allowed the price of a pound of dry parchment coffee to increase by up to 40-fold (when coffee is sold by the cup), thus improving the living conditions of the families involved in the process.

Garittea is also designed as a living laboratory that brings the reality of the countryside closer to the city and the city closer to the countryside. In this respect, Garittea enables more direct communication to take place between producers and consumers and opens spaces for the exchange of knowledge and experiences between peasant and academic communities.

The exchanges and transactions are not exclusively monetary. There are spaces for the exchange of knowledge in which producers offer lectures on agroecology and healthy food, as well as training on quality and methods of coffee preparation. The academic community tackles other forms of economy management, the systematizing of processes, and the contribution from their areas of knowledge to the

[5] Association of Organic Coffee Growers of Colombia (ACOC) and the Association of Coffee Growers of the township of La Marina (ASOPECAM)

[6] Instituto Mayor Campesino (Social Center of the Comparte Network)

development of a more aware and supportive university community around Garittea. Among the academic activities are subjects, degree theses, student internships, and systematizations, including the book "CONVITE: systematization of a collaborative experience in which the challenges and tensions of production-creation are reflected in a collaborative environment based on the Garittea process" (Mora-Anto, F. et al. 2018).

In 2023, the turnover of the Capeltic coffee shops amounted to $840,000 USD, thus improving the direct income of more than 280 indigenous producer families who, through their coffee production, conserve 236 Ha of tropical forests. In the same way, by processing more than a ton of agroecological coffee in cups, the turnover of Garittea amounted to $80,000 USD, and there was an improvement in the income of 25 families who conserve more than 50 Ha of tropical forests through their agroecological practices.

The inter-cooperation experience between Capeltic and Garittea mediated by the Comparte Network is mainly due to the following three characteristics of the Network:

(a) **The Learning/Action Community Character of the Comparte Network**: The exchange of experiences and the generation of transdisciplinary knowledge is the essence of the Network's work. Thus, in 2004, through the peasant-to-peasant methodology mediated by the Bachajón Mission and the Instituto Mayor Campesino (IMCA), a cooperative relationship was initiated between rural indigenous people from Chiapas, Mexico and organic coffee farmers from the Valle del Cauca, Colombia. These exchanges led to the production and marketing of organic coffee by Bachajon Mission in Chiapas, Mexico. This proces began with a small-scale pilot group that, over the years, grew into a group of social enterprises that continue to implement regenerative agroecological practices. The result: roasted coffee, sold through its Capeltic coffee shops in alliance with the Universidad Iberoamericana. Six years later representatives of the IMCA and the Javeriana University visited Yomol A'tel and the Capeltic coffee shops in Mexico. In this experience they saw a way not only of improving their local processes, but also of making themselves visible as a production initiative with the capacity to control the different links in the coffee value chain. In this respect, just as the producers of Valle del Cauca inspired the development of the organic coffee value chain project in Mexico, Capeltic became a goal and support for the formation and development of Garittea in Colombia.

(b) **The Commitment to the Multi-actor Approach:** Within the framework of collaboration between social centers and universities, the Office of University Social Responsibility of the Pontificia Universidad Javeriana (PUJ) promoted a scheme enabling students mainly from urban contexts to learn about the processes of territorial sustainability in rural areas through the Instituto Mayor Campesino. At the University, a group of students promoted by this office came up with the idea of marketing products brought directly from the field.

This idea is beginning to gain strength at the university and is added to the results of the exchange of experiences between Capeltic, the IMCA, and the University teaching staff. This resulted in the formal incorporation of the Javeriana University into the Garittea co-creation project, in which around 400 students and 22 professors from various faculties, such as design, architecture, communication, human sciences, and administration, joined the process. The same thing had taken place years earlier in the case of Capeltic, where it was professors and students from the Universidad Iberoamericana CDMX who, hand in hand with the producers, promoted the development process of the value chain and the incorporation of the coffee produced into international markets.

(c) **Capacity Building Through the Implementation of Methodologies Developed Within the Network**: The *methodology route for the construction of economic-production alternatives* was the basis for identifying the multidimensional potentialities and challenges of this project, as well as identifying the key actors. The knowledge gathered through this methodology was combined with that of the academy, the strategic framework, and social business plan, and other action plans for the different areas were thus developed.

A multi-stakeholder approach was developed in the composition of both social enterprises. For example, the co-creation of Garittea, inaugurated in 2017, required the synergy of various actors, such as rural associations, NGOs, universities, companies, and volunteers who contributed from their own talents and knowledge (Mora-Anto, F. 2018). The first step was the mediation of the IMCA in the rapprochement between actors and the recognition that each of them was a fundamental part of the success of the process, thus adding directionality to their commitments and efforts for a common project.[7]

In addition, the agroecological coffee production process has been linked to the *Participatory System of Guarantees (SPG)*, which generates a direct bridge between those who produce and those who consume, without depending on external certifications. That, in turn, allows it to expand its marketing horizons through local markets in the region, such as the Network of Agroecological Markets of the Valle del Cauca.

Likewise, other ongoing processes are added to these processes. These include the *solidarity economic circuits (CES)* that coordinate different expressions of the social and solidarity economy in a specific territory, and that seek to satisfy the fundamental needs of local communities and generate a culture of *Buen vivir*.

[7] This process of rapprochement between different actors, mainly between rural organizations and the university, led to encounters and disagreements that are explained in the systematization book CONVITE.

7.4 Reflections and Implications

The experience of the Comparte Network shows how hope can be built and sustained in complex contexts such as Latin America. Through its territorial perspective and multi-stakeholder strategy, Comparte has been supporting the transformation of territories with a transdisciplinary approach to learning and action that seeks to strengthen sustainable economic alternatives that generate a territorial impact.

Lasting hope is built from the territory by strategically orienting its potentialities in terms of collective decolonization projects. The *Lectura Estratégica del Territorio* (LET, Spanish for Strategic Reading of the Territoty) is a management planning exercise that involves key stakeholders with a clear conceptual and practical understanding of *Buen vivir*.

Integral ecology and the *Buen vivir* paradigm open up new horizons of possibilities and strengthen the resilience of people who, along with the organizations they form, constantly seek to earn a living through a relational paradigm that prioritizes the care of life and harmony with all beings in nature, all the while subordinating the economy to this end.

The methodology proposed in this chapter is based on a territorial strategy that articulates the expressions of a region's alternative economies. This strategy is realized through the strengthening of collective initiatives based on the potential that results in the social redistribution of the wealth generated. The experiences of Capeltic and Garittea offer us a horizon of action that exemplifies this path, which the Comparte Network, inspired by Pope Francis's words, embraces with hope:

> Christian hope is having the certainty that I am on the path to something that is, and not to what I want it to be.[8]

The dynamics of inter-cooperation generated by the Network provide people and organizations with capacities and instruments. They also strengthen and consolidate their commitment to the construction of an economy that is in service to life. For instance, this perception is evident in the testimonies of the people who have participated in the meetings and methodology processes of the Network with expressions such as "we are not alone."

The dynamics of inter-cooperation also confirm that lasting hope emerges from collective experiences (communities, organizations, networks, and ecosystems) that lead to actions (methodologies, experiences) capable of materializing the transition from theological hope to a transformative practice.

Hope emerges at the local level and scales up to the global in the same way that the coffee bean is produced in Latin American agroecological soils and is enjoyed in a cup in Colombia, Switzerland, Germany, and elsewhere. Thus, alternative ways of organizing life and the economy emerge—rooted in solidarity, care, and sustainability—which are shared and enriched through collective learning, and can be adapted and replicated across diverse contexts.

---

[8] GENERAL AUDIENCE, Paul VI Audience Hall, Wednesday, 1 February 2017

Specific scalable and sustainable paths and horizons are needed to advance toward experiences that offer an alternative to development models based solely on economic growth. The challenge is to find, test, and systematize the forms and tools that allow us to move from local and micro-scale experiences to subregional or regional experiences that make an impact. These experiences must have increasingly broader perspectives within the indispensable framework of food sovereignty, the political and cultural autonomy of communities, and correct agroecological practices that can be replicated in other contexts.

References

Aguilar JA, Idárraga A (2017) Ruta metodológica para la construcción de alternativas económico-productivas de impacto local y regional. Pontificia Universidad Javeriana de Cali, Colombia

Altmann P (2017) Sumak Kawsay as an element of local decolonization in Ecuador. Lat Am Res Rev 52(5):749–759

Borda-Rodriguez A, Johnson H, Shaw L, Vicari S (2016) What makes rural cooperatives resilient in developing countries? J Int Dev 28(1):89–111. https://doi.org/10.1002/jid.3125

Deane-Drummond, C. (2023). Pope Francis's integral ecology paradigm: an exploration of its theological foundations and ethical implications. Ir Theol Q 88(2):99–111.

Deane-Drummond C, Deneulin S (2021) Integral ecology: discovering its meaning and depth'. LSRI briefing note 1. Oxford: Laudato Si. Research Institute, Campion Hall

Esteves A (2021) Processes of normative regulation in spaces of "solidarity economy": a comparative case study analysis. Int J Sociol Soc Policy. https://doi.org/10.1108/IJSSP-12-2020-0540

Fuentes A (2023) Alternativas en tiempos de crisis civilizatoria. Editorial Universidad, Iberoamericana

Francisco. (2015). Laudato Sí. *Carta encíclica.* Sobre el cuidado de la casa común. Editorial Vaticano. https://www.vatican.va/content/francesco/en/encyclicals/documents/papa-francesco_20150524_enciclica-laudato-si.html

Hillebrand, Zademach (2013). Alternative economies and spaces: introductory remarks. *Altern Econ Spaces* ,9–22

Irezabal, A. (2023). "Bienestar Comunitario. El Buen Vivir." *Revista IBERO. Calidad de Vida y Desarrollo Humano en México.* Universidad Iberoamericana. México. ISSN 25942662. https://revistas.ibero.mx/ibero/uploads/volumenes/73/pdf/6-bienestar-comunitario-el-buen-vivir-alberto-irezabal-vilaclara.pdf

Kim D, Cho W, Allen B (2020) Sustainability of social economy organizations (SEOs): an analysis of the conditions for surviving and thriving. Soc Sci J (Fort Collins):1–17. https://doi.org/10.1080/03623319.2020.1799174

Magallón M., López M., & Zepeda C. (2022). Evaluación de la coherencia interna del modelo de intervención de la red Comparte en América Latina. https://drive.googlsssse.com/file/d/1Ph8Q7oMA0Ktjv9BUf7Lm5KBcjnCkFgH/view?usp=sharing

Merçon, J. (2022). Investigación transdisciplinaria e investigación-acción participativa en clave decolonial. *Utopía y Praxis Latinoamericana,* 27(98):e6614174

Mora-Anto, F. (2018) *Convite. Sistematización de una experiencia colaborativa.* Pontificia Universidad Javeriana. Cali, Colombia

Nicolescu, B. (1996). *La transdisciplinariedad. Manifiesto.* Multiversidad Mundo Real Edgar Morin, A.C. México

Parida, Wincent. (2019). Why and how to compete through sustainability: a review and outline of trends influencing firm and network-level transformation. *Intern Entrepreneurship Management J* 15(1):1–19. Springer

RCS-CPAL. (2022). Guidance framework for studying and working with integral ecology. *Conference of Provincials from Latin America and the Caribbean (RCSCPAL).* https://jesuitas.lat/

Diana Patricia Giraldo PhD candidate in Development Studies at the University of Lisbon and a member of the technical team of the Comparte Network. She is an economist with a master's degree in Globalization and Development from the University of the Basque Country (Spain). Her professional trajectory combines over two decades of experience in the retail sector and international cooperation. Her research focuses on the transformative potential of Solidarity Economic Circuits, mechanisms of socio-economic change, and the construction of territorial alternatives grounded in Buen Vivir.

Alberto Irezabal Director of the International Center for Research on the Social and Solidarity Economy at the Ibero-American University, Mexico City. Trained as an industrial engineer and holding a PhD in Advanced Management of Organizations and Social Economy, his work focuses on social and solidarity economy, community-based development, and sustainable value chains. Drawing on more than a decade of collaboration with Tseltal Indigenous communities in Chiapas, he explores pathways for fostering inclusive economic development.

María del Mar Magallón PhD candidate in Business and Territorial Competitiveness, Innovation and Sustainability at Deusto Business School, Spain, and a member of the Steering Committee of the Comparte Network. Drawing on 20 years of experience in the management and leadership of international development organizations, her research interests focus on participatory evaluation, impact measurement and capacity strengthening in contexts of exclusion.

Álvaro Idárraga An agronomist, agroecology specialist, and educator. He is the former Executive Secretary of the COMPARTE Network and has extensive experience in organic agriculture, rural development, and collaboration with peasant organizations in Colombia.

Erminsu David An agronomist trained at the National University of Colombia holding a PhD in Social Economy. He serves as Social Promotion Coordinator at the Instituto Mayor Campesino (IMCA) and currently coordinates the COMPARTE Network in Latin America. His work focuses on agroecology, peasant economies, and rural development, drawing on extensive experience working with farming communities.

Chapter 8
Regenerative and Climate-Smart Agriculture at High Altitude for Sustainable Development

Empowering Mountain Communities Through Saffron Cultivation in the Hindu Kush, Pakistan

Dinar Wali and Peter Marty

Abstract Mountain regions face increasing environmental and socioeconomic challenges that require locally adapted pathways for sustainable development. This chapter examines a climate-smart and regenerative agriculture initiative in Upper Chitral, Pakistan, where saffron cultivation was introduced as a high-value crop suited to the ecological conditions of the Hindu Kush. The regional context is characterized by limited arable land, climatic vulnerability, and constrained economic opportunities, creating a strong need for resilient agricultural systems. The chapter describes the community-based approach of RegenX.eco, emphasizing stakeholder collaboration, women's participation, capacity building, and market-oriented value creation. A pilot project was implemented in five villages selected according to altitude, soil suitability, accessibility, and community engagement. Soil analyses and environmental assessments informed cultivation practices based on organic amendments, efficient water management, and regenerative farming principles. The results demonstrated successful establishment of saffron under high-altitude conditions, with strong germination, flowering, and promising quality characteristics. Beyond agronomic outcomes, the initiative generated social benefits through women's empowerment and additional income opportunities while contributing to soil health, biodiversity conservation, and climate resilience. The chapter further highlights prospects for value addition, product diversification, and integration into premium markets, while identifying challenges related to infrastructure, bulb supply, and institutional support. The findings indicate that saffron cultivation can provide a viable and scalable model for sustainable regional development in mountain areas.

D. Wali
regenx.eco, Islamabad, Pakistan
e-mail: dinar.wali4@gmail.com

P. Marty (✉)
Zurich University of Applied Sciences ZHAW, Wädenswil, Switzerland
e-mail: peter.marty@zhaw.ch

T. Karcher SJ et al. (eds.), *Making Hope Possible*, Environmental Humanities: Transformation, Governance, Ethics, Law 7,
https://doi.org/10.1007/978-3-032-17719-3_8

Keywords Regenerative agriculture · Climate smart farming · Farming in high altitudes · Saffron cultivation · Mountain development · Pakistan · Community-based development

8.1 Introduction

8.1.1 About This Paper

Agriculture continues to be a cornerstone of human civilization, providing sustenance and livelihoods while forming a basis for economic development worldwide. In mountainous regions such as the Chitral Valley in northwestern Pakistan's Hindu Kush range, agriculture takes on added significance due to the limited arable land available, harsh climatic conditions, and the increasing vulnerability of local communities to the impacts of climate change. As global awareness of environmental sustainability and climate resilience grows, the demand for climate-smart and regenerative agricultural practices intensifies. Within this broader context, RegenX.eco[1] promotes innovative, collaborative, and sustainable interventions aimed at bolstering the livelihoods of communities in Chitral by cultivating high-value crops such as saffron.

This paper presents a comprehensive overview of the saffron cultivation pilot project launched by RegenX.eco in Upper Chitral. It begins by placing the project in its geographical, socioeconomic, and climatic contexts, exploring the factors that have historically constrained agricultural development in this high-altitude environment. The discussion then turns to the rationale for choosing saffron (Crocus sativus), examining its botanical attributes, global market dynamics, and the health and economic benefits that make it an attractive crop. Following a detailed presentation of soil data, cultivation practices, and pilot results, the paper delves into saffron's broader environmental and economic implications. It outlines pathways for scaling and sustainability. Ultimately, the findings emphasize the value of multi-stakeholder collaboration, the importance of tailoring agricultural interventions to local contexts, and the long-term potential of saffron cultivation as part of a climate-smart regenerative agriculture model in high-altitude regions, thereby contributing to sustainable development.

---

[1] https://regenx.eco/.

8.1.2 Concept of Sustainable Development in Mountain Regions

"Mountains are key contexts for sustainable development because they provide indispensable goods and services. However, they are among the most disadvantaged regions in the world, with the highest poverty rates and some of the greatest vulnerability to global climatic, environmental, and socioeconomic change and related risks" (Wehrli 2014). The 2030 Agenda aims to enable "a better and more sustainable future for all." In the context of the 2030 Agenda, the importance of mountain regions for sustainable development is highlighted through several specific targets. These include the protection and restoration of water-related ecosystems such as mountains, forests, wetlands, rivers, aquifers, and lakes (SDG 6.6); the conservation, restoration, and sustainable use of terrestrial and inland freshwater ecosystems and their services, with particular attention to forests, wetlands, mountains, and drylands (SDG 15.1); and the conservation of mountain ecosystems, including their biodiversity, in order to strengthen their capacity to provide essential benefits for sustainable development (SDG 15.4).

The 2030 Agenda does not explicitly emphasize the needs and priorities of the mountain population. This makes it all the more important to adopt an approach that considers a region's socioeconomic and sociocultural conditions, identifies its inherent potentials, and leverages them effectively to value.

8.2 Background and Context of the Case Study

8.2.1 Geography of Chitral Valley and the Hindu Kush Mountains

Chitral (Fig. 8.1) is located in Pakistan's Khyber Pakhtunkhwa (KP) province, nestled within the Hindu Kush mountain range. Known for its majestic peaks, glaciers, and winding rivers, Chitral boasts breathtaking landscapes but also faces geographical constraints that complicate traditional agriculture. Rugged mountains dominate most of the region; only about 10% of the 14,850 km^2 is arable, concentrated in river valleys and terraces. The district's elevation varies from 3700 ft (1130 m) in the lowest valleys to over 25,000 ft (7620 m) at mountain summits. For saffron cultivation, altitudes ranging from 6400 ft (1950 m) to 8000 ft (2440 m) are particularly important for evaluating microclimatic suitability, soil composition, and potential growing seasons.

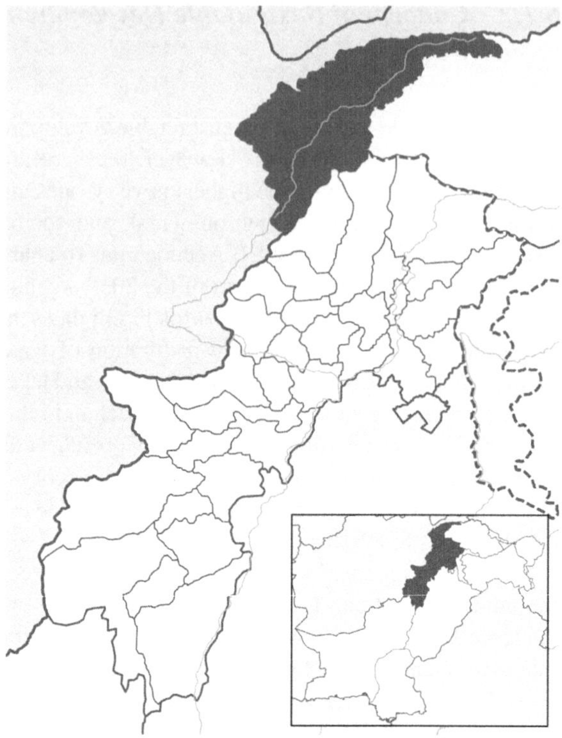

Fig. 8.1 The Upper Chitral District in the province of Khyber Pakhtunkhwa is located in the northwest of Pakistan, on the border with Afghanistan, in the Hindu Kush mountain range. It has a population of 195,528 inhabitants (Wikipedia 2024)

8.2.2 Socioeconomic Context and Agricultural Heritage

Historically, the people of Chitral have relied on subsistence farming, pastoralism, and small-scale trade. Barley, wheat, and maize are the staple crops, while fruit orchards producing apples, apricots, and walnuts are found at lower altitudes. Livestock rearing—particularly sheep, goats, and cattle—adds diversity to household economic activities. However, rapid socioeconomic changes, limited market access, and inadequate infrastructure have constrained farmers' capacity to transition from subsistence to commercial farming. As a result, many families rely heavily on remittances from labor migration within Pakistan and abroad.

8.2.3 Climate Change Vulnerability

Climate change remains a major challenge in the Hindu Kush region. According to various studies, glaciers in the Hindu Kush–Karakoram–Himalaya system are retreating at an alarming rate (Bolch et al. 2012; Lutz et al. 2014). Alterations in temperature and precipitation patterns threaten water availability, agricultural productivity, and ecosystem services. Glacial lake outburst floods (GLOFs) and unpredictable monsoon patterns further heighten the risk of natural disasters such as rockfalls, floods that affect infrastructure and settlement area, and landslides. In light of these challenges, communities in Chitral face a precarious future, necessitating the adoption of climate-resilient farming methods that enhance productivity, conserve biodiversity, and secure livelihoods in the face of environmental uncertainties.

8.3 RegenX.eco: A Social Entrepreneurship Venture

8.3.1 Vision and Objectives

RegenX.eco is a social enterprise that promotes Climate-Smart Regenerative Agriculture in the high-altitude Pamir region of the Chitral Valley in northwestern Pakistan. Rooted in the principles of sustainability, social equity, and value addition, the initiative envisions a farming system that improves soil health, conserves water resources, and uplifts local communities—particularly women—through skill-building and income generation.

8.3.2 Approach to Community Collaboration

RegenX.eco's engagement model is based on a collaborative and participatory approach that combines scientific knowledge with local experience. Through Participatory Rural Appraisal (PRA), the initiative identifies community needs, available resources, and local expectations in order to ensure that interventions are aligned with local priorities. Capacity-building activities, particularly for women, focus on crop selection, organic farming practices, and processing techniques. In parallel, value-chain development efforts seek to identify local, regional, and international market opportunities for specialty crops such as saffron, enabling farmers to access premium market segments and increase household incomes. Regular monitoring and evaluation activities support project implementation by tracking progress, assessing impacts, and informing the continuous refinement of interventions.

8.3.3 Prior Initiatives and Partnerships

Before focusing on saffron cultivation RegenX.eco experimented with various high-value horticultural products and eco-tourism initiatives. In 2018–2020, a successful saffron cultivation project in the Western Himalayan region of Uttarakhand and Himachal Pradesh, India (Kothari et al. 2021), provided a model for introducing saffron cultivation in Chitral. Meanwhile, partnerships with North Naturals[2] and entities in the United States have been established to facilitate product development and to access new markets for saffron-based pharmaceuticals, cosmetics, and high-end culinary products.

8.4 Why Saffron (*Crocus Sativus*)?

8.4.1 Botanical Characteristics and Suitability

Saffron is derived from the stigmas of the autumn-flowering perennial crocus (*Crocus sativus*), a member of the Iridaceae family. Each saffron flower bears three stigmas, which, once carefully harvested and dried, become the saffron threads or strands. The crop thrives in regions with well-drained, loamy or sandy-loam soils. It can withstand drastic diurnal temperature fluctuations, conditions characteristic of many arid and semi-arid, high-altitude landscapes.

Saffron possesses several agronomic characteristics that make it particularly suitable for cultivation in high-altitude environments. The crop is highly drought tolerant and requires relatively little water compared with many conventional agricultural crops. At the same time, saffron offers exceptional economic value, ranking among the most expensive spices in the world on a per-kilogram basis. Its cultivation cycle is also well aligned with the seasonal conditions of many mountain regions, with planting taking place from late summer to early autumn and flowering occurring in late autumn, thereby fitting into agricultural windows that are often underutilized in high-altitude farming systems.

8.4.2 Historical Cultivation and Global Production

Historically, saffron has been cultivated in Iran, Spain, India (particularly in Kashmir, Himachal Pradesh, Uttarakhand), Greece, and Morocco (Fernández 2004). Iran remains the largest producer, accounting for more than 90% of global output (FAO 2021). However, shifting climatic patterns and the adoption of more profitable cropping systems have raised concerns about the future of global saffron supply. This situation creates market opportunities for new producers such as Chitral.

[2] https://northnaturals.com/.

8.4.3 Health Benefits and Medicinal Properties

Saffron has long been valued for its potential health-promoting properties and bioactive compounds. It contains crocin, crocetin, picrocrocin, and safranal, which are associated with antioxidant and anti-inflammatory activities. Research has also suggested that saffron may contribute to mood regulation and may help alleviate symptoms of mild to moderate depression. In addition, preliminary studies have indicated potential anticancer effects through the inhibition of tumor development, while further evidence points to possible protective benefits for cardiovascular health. These diverse properties contribute to saffron's growing importance in nutraceutical, pharmaceutical, and wellness-oriented applications.

8.4.4 Market Analysis and Economic Importance

The global saffron market has grown steadily driven by increasing demand from the culinary, pharmaceutical, and cosmetic sectors. According to a report by Grand View Research (2020), the global saffron market size is expected to continue expanding at a steady compound annual growth rate (CAGR) in the coming years. High demand in regions such as the Middle East, Europe, and North America—combined with constrained global supply—has sustained saffron's premium market price. As a result, saffron cultivation is particularly attractive for smallholder farmers in mountainous regions where alternative high-value crops are scarce.

8.5 Pilot Project in Upper Chitral

8.5.1 Project Rationale and Objectives

RegenX.eco identified saffron as a promising crop for Upper Chitral based on a combination of ecological, agronomic, economic, and resilience-related factors. The region shares important climatic and environmental characteristics with other saffron-producing mountain areas, particularly in parts of the Himalayan belt. Local soils, characterized by loamy textures and beneficial organic matter content, provide favorable conditions for saffron cultivation. In addition, the crop's exceptionally high market value offers significant opportunities for income generation and rural economic development. Its drought tolerance further enhances its suitability for the water-scarce environments typical of high-altitude regions, making saffron a potentially resilient and sustainable agricultural alternative for local communities.

Therefore, a 2023 pilot project aimed to test the feasibility of saffron production, assess community interest, and build local capacity for regenerative, climate-smart agricultural methods.

Table 8.1 The five villages of the pilot phase in 2023 in Upper Chitral with a brief description

Village	Altitude	Total households	Climatic conditions	Soil quality
Charun	6400 ft 1,950 m	450	Dry, low humidity, primarily sunny	Fertile brownish soil
Junali Koch	6500 ft 1,980	350		Sandy loam soil texture
Booni	6500 ft 1,980 m	2,000	Cold weather because of glacier	Brown soil texture
Mastuj	7290 2,200 m	1,100		Sandy loam soil texture
Chapali	8000 2,440 m	400	Cold and dry weather	Sandy loam soil texture

8.5.2 *Site Selection Parameters*

The selection of the five pilot villages in Upper Chitral was guided by a combination of agronomic, environmental, social, and logistical criteria. Villages located between 6,400 and 8,000 ft (1,950–2,440 m) were chosen to evaluate saffron performance under different microclimatic conditions and altitude gradients. Particular attention was paid to soil suitability, with priority given to fields exhibiting balanced pH levels and adequate organic matter content, both of which are important factors for successful saffron cultivation. Community engagement also played a central role in the selection process. Preference was therefore given to locations characterized by strong local leadership and a high level of interest among women willing to participate in cultivation activities. In addition, practical considerations such as road access and the availability of suitable irrigation infrastructure were taken into account to facilitate project implementation and effective monitoring throughout the pilot phase. The five villages selected for the 2023 pilot project are presented in Table 8.1.

8.5.3 *Timeline of Activities*

The cultivation activities were carried out with the active participation of women from Local Support Organizations (LSOs). These have been established in Upper Chitral since the 1990s by the Aga Khan Rural Support Programme (AKRSP; Aziz Ali Dad 2016). Twenty women were involved in saffron bulb planting and cultivation per acre of land. This number varies during the harvesting and field-clearing stages. Ideally, 11–14 women growers are required per acre of farmland.

The implementation of the pilot project followed a clearly defined timeline. Between 13 and 23 September 2023, RegenX.eco, in collaboration with Rapid Solution, imported and distributed 30,000 saffron bulbs across five villages in Upper Chitral. Initial sprouting and vegetative growth were observed in late October, followed by flowering and stigma harvesting in early November. As the pilot sites differed in altitude, weather conditions, soil characteristics, and organic matter content, variations in sprouting and flowering dates were recorded among the villages. In mid-November, harvested saffron underwent drying and quality assessment. Throughout the pilot phase, RegenX.eco worked closely with local growers to facilitate knowledge transfer and build technical capacity. While cultivation activities were jointly managed during the initial stage, the long-term model envisages community-based saffron production supported by RegenX.eco through value addition, marketing, and market access services.

8.5.4 Training and Capacity Building

To support the successful establishment of saffron cultivation, RegenX.eco organized a series of training activities for local growers. These focused on the inspection and selection of healthy bulbs, appropriate planting depths, and the application of organic soil enrichment techniques to improve growing conditions. Participants also received guidance on irrigation scheduling tailored to local environmental conditions, as well as training in harvesting methods, safe handling practices, and post-harvest procedures. Particular emphasis was placed on drying and storage techniques to preserve saffron quality and maximize its market value. Together, these capacity-building measures were designed to strengthen local expertise and enable growers to manage saffron production independently over time.

The overarching objective was to ensure local growers to independently manage production after the pilot phase, thereby fostering a sustainable model for scaling saffron cultivation.

RegenX.eco implemented the project in partnership with local growers. At scale, growers will manage production independently, while in the initial phase technical guidance, training, and knowledge transfer are essential. Therefore, capacity building and skills development for local farmers are crucial. Ultimately, the growers will produce saffron independently, while RegenX.eco will manage value addition, export, and contract purchasing.

8.6 Soil and Environmental Analysis: Five Villages

8.6.1 Overview of Data Collection

Soil sampling, temperature monitoring, and altitude measurements were in September 2023. Each village site was analyzed for soil texture pH, organic matter, nitrate (a proxy for nitrogen), phosphorus, potassium, and electrical conductivity. These laboratory analyses were conducted by RegenX.eco on the soils.

The project team took samples from five villages to determine soil macronutrient levels. Differences in these parameters informed irrigation schedules and fertilizer recommendations, ensuring optimal growing conditions for saffron bulbs.

8.6.2 *Detailed Village Profiles with Soil Data*

A consolidated list of the five pilot villages with their respective data points can be found in Table 8.1.

8.6.3 *Comparison and Discussion of Soil Data*

Across the five pilot sites, pH values ranged from 6.70 in Junali Koch to 7.59 in Charun, a range well-suited to saffron, which thrives in soils with a pH between 6.0 and 8.0. Organic matter content varied significantly, from 1.55 g/kg in Junali Koch to around 3.30 g/kg in Charun. Higher organic matter improves moisture retention and nutrient availability, which are critical for saffron bulbs during the initial rooting stages. Nitrate (N) levels were generally low, which is of mountainous soils, but still within a range where additional organic amendments could enhance productivity. Phosphorus (P), an essential macronutrient for root development, ranged from 2.20 to 7.95 g/kg. This wide variation indicates the need for site-specific fertilization. Potassium (K) values were relatively high, ranging from 168 to 250 g/kg, supporting overall plant vigor, although interpretation depends on the chemical form of potassium in the soil. Lastly, electrical conductivity (EC) remained below 1 dS/m in most samples, indicating low salinity, which is favorable for saffron.

8.7 Saffron Cultivation Practices

8.7.1 *Land Preparation and Bulb Planting*

Land preparation began in early September 2023 with tilling fields to a depth of approximately 20–25 cm. Guided by the soil analysis results, organic compost and well-rotted manure were added where required. Bulbs (corms) imported from Iran were carefully inspected for signs of disease and damage. Planting occurred from September 13 to September 23, 2023, at a spacing of 10–15 cm between bulbs and a planting depth of about 10 cm.

During planting careful control of bulb depth and spacing was essential. The reason is that saffron bulbs typically multiply by a factor of approximately three each year, and the bulbs remain in the soil for three years. Adequate spacing between corms, together with proper air circulation and sufficient nutrient and water availability, is thus essential for effective bulb regeneration.

8.7.2 Irrigation Regimes and Water Management

One significant advantage of saffron is its relatively low water requirement, especially after the bulbs have become established. The pilot project used drip or flood irrigation at intervals of 14–17 days, depending on measured soil moisture levels. Careful timing was crucial to prevent overwatering, which can lead to bulb rot in cooler mountain climates. Despite water scarcity in many parts of Chitral, saffron's drought tolerance makes it a viable option.

8.7.3 Organic Soil Amendments and Pest Control

RegenX.eco applied organic farming practices, avoiding synthetic pesticides and chemical fertilizers. Crop residues, compost, and farmyard manure were used to improve soil structure and nutrient density. Occasional foliar sprays of neem-based or garlic-based biopesticides helped mitigate common fungal and insect threats. Rotating saffron with other low-water, high-value crops (such as certain medicinal herbs) can break pest cycles and maintain soil fertility.

8.7.4 Harvesting, Drying, and Post-harvest Handling

Saffron flowers began blooming in early November 2023, with flowering lasting approximately one week. Flower coloration exhibited variations related to altitude: deeper purple hues at lower elevations (Charun, Junali Koch) and lighter pinkish hues at higher elevations (Chapali, Mastuj). Harvesting was carried out manually, usually early in the morning, to preserve aroma and potency. Women carefully collected the flowers, hand-plucked the stigmas, and transferred them to well-lit and well-ventilated drying areas. Sun drying reduced stigma weight, producing the characteristic thin saffron threads with high market value. Visual documentation of the cultivation process is presented in Figs. 8.2, 8.3, 8.4, 8.5 and 8.6.

Fig. 8.2 Village Charun 6400 ft in the Hindu Kush Range, Chitral, Pakistan. The image quality of the photograph was enhanced using AI

Fig. 8.3 Saffron bulbs imported from Iran for the proof of concept in Fall 2023 (©Dinar Wali 2023). The image quality of the photograph was enhanced using AI

Fig. 8.4 Women growers seeding saffron bulbs in village Charun on September 13, 2023 (©Dinar Wali 2023). The image quality of the photograph was enhanced using AI

Fig. 8.5 Full bloomed saffron flower in village Charun, Upper Chitral, Hindu Kush Range (©Dinar Wali 2023). The image quality of the photograph was enhanced using AI

Fig. 8.6 Dried saffron threads (©Dinar Wali 2023). The image quality of the photograph was enhanced using AI

8.8 Pilot Project Outcomes

8.8.1 Germination, Blooming, and Flower Colour Variation

The pilot project produced encouraging agronomic results across all five villages. Germination rates were consistently high, reflecting both the quality of the imported Iranian corms and the suitability of local soil conditions for saffron cultivation. Flowering commenced during the first week of November 2023 and continued for approximately seven to ten days. Variations in flower coloration were observed along the altitudinal gradient, ranging from deep purple at elevations of approximately 6,400–6,500 ft to lighter pinkish-purple hues at around 8,000 ft. While these differences did not appear to influence saffron quality, they suggest the importance of local microclimatic conditions in shaping phenotypic characteristics and highlight an area for further scientific investigation.

8.8.2 Yield and Quality Assessment

Although final yield data is still under evaluation, initial indicators suggest that yields could be significantly improved through better soil fertility and water management. Preliminary quality assessments indicate that saffron threads possess robust color (dark red) and intense aroma, indicating the potential for Category I classification under ISO 3632 standards (the international standard for saffron quality).

8.8.3 *Socioeconomic Impact on Women Growers*

One of the core objectives of RegenX.eco was to empower women through their involvement in every stage of saffron cultivation. Each acre required approximately 10–12 women to carry out planting, weeding, harvesting, and drying activities. The additional income generated has expanded household earnings and enhanced women's confidence and social standing within their communities. Many participants expressed interest in expanding their saffron plots in the coming seasons.

8.9 Environmental and Sustainability Considerations

8.9.1 *Climate-Smart Agriculture Principles*

Saffron cultivation in Upper Chitral demonstrates a strong alignment with the core principles of climate-smart agriculture. Its exceptionally high market value enables farmers to generate substantial returns from limited land resources, with dried saffron yields increasing from approximately 300 g per acre in the first year to 700 g in the second year and 1 kg in the third year. At the same time, the crop's low water requirements and ability to perform well in marginal soils enhance the resilience of farming systems and reduce vulnerability to climate-related stresses. Furthermore, the reliance on organic management practices, coupled with the maintenance of soil organic matter, contributes to climate change mitigation by supporting carbon sequestration and improving overall soil health.

8.9.2 *Carbon Sequestration and Soil Health*

By prioritizing organic amendments, minimal tillage, and diversified cropping systems, saffron cultivation can incrementally boost soil organic carbon (SOC) levels. Healthy soils with higher SOC function as more effective carbon sinks, thereby contributing to climate change mitigation at the local level. Over extended periods, these regenerative practices may yield ecosystem benefits such as improved water retention and reduced erosion—critical in mountainous terrains prone to landslides.

8.9.3 *Biodiversity Preservation*

Unlike monocultural plantations, saffron plots in Chitral often coexist with traditional crops, fruit orchards, and surrounding natural vegetation. Intercropping or rotating saffron with local legumes and other cover crops can enhance on-farm biodiversity, offering habitats for beneficial insects, pollinators, and microorganisms that contribute to a balanced agroecosystem.

8.10 Economic Analysis and Value Addition

8.10.1 Local Value Chains and Market Channels

Currently, saffron produced under the pilot project is sold at premium prices in local markets. However, production volumes are not yet sufficient to attract large-scale buyers. The next step involves establishing direct linkages with national spice in Pakistan and exploring e-commerce platforms to reach consumers seeking organic, ethically sourced saffron.

8.10.2 Partnership with North Naturals

RegenX.eco's partnership with North Naturals has been an important component of the project. North Naturals develops and markets saffron-infused products, ranging from wellness teas to speciality cosmetics. Their global market reach can position saffron from Chitral in premium consumer markets in the Middle East, the United States, Europe, China, Japan, and Korea. The collaboration also enables local farmers to access advanced processing technologies and achieve higher margins by ascending the value chain.

8.10.3 Global Market Opportunities (MENA, US, EU, China, Japan, Korea)

Demand for saffron in Middle Eastern and North African (MENA) markets remains strong due to cultural and culinary traditions favoring saffron. Europe and the United States have experienced growing interest in saffron-based nutraceuticals and gourmet products. Meanwhile, the rising middle class in China and increasing demand for holistic wellness solutions in Japan and Korea present promising emerging markets. Strategic marketing—emphasizing saffron's organic provenance, fair-trade value, and traceability—can help saffron from Chitral secure a niche in these competitive markets (Grand View Research 2020).

8.10.4 *Product Diversification (Pharmaceutical, Cosmetic, Culinary)*

Saffron's potential extends beyond conventional culinary use (in dishes like biryani and paella). Pharmaceutical research focuses on saffron's bioactive compounds, with potential applications in mood enhancement, anti-inflammatory therapies, and oxidative stress management (Razavi and Hosseinzadeh 2017). In cosmetics, saffron extracts are used as natural skin-brightening and antioxidant ingredients. Food product innovations include saffron-infused honey, syrups, cheeses, and premium teas. By diversifying into these market segments smallholder farmers in Chitral can capture more value from each ounce of saffron produced.

8.11 Challenges, Lessons Learned, and Future Prospects

8.11.1 *Limitations in Scaling and Infrastructure*

Despite the encouraging results of the pilot project, several challenges must be addressed to support the long-term scaling of saffron cultivation in Upper Chitral. Continued dependence on imported corms may increase production costs and create supply constraints, highlighting the need for locally established corm nurseries to ensure a reliable and affordable source of planting material. Although saffron requires relatively little water, consistent irrigation management remains essential for achieving stable yields, making investments in improved water distribution systems and efficient drip irrigation technologies desirable. In addition, adequate post-harvest handling is critical for maintaining product quality and market competitiveness. Expanding drying, sorting, and packaging facilities, together with targeted training in post-harvest management, will therefore be important for strengthening the value chain and supporting the sustainable growth of saffron production in the region.

8.11.2 *Policy and Institutional Support*

Government recognition of saffron as a strategic cash crop in high-altitude regions could lead to subsidies for bulbs, tax incentives, and infrastructure investments. Engagement with local agricultural departments and the Khyber Pakhtunkhwa government could help ensure that saffron becomes an integral part of regional development plans.

8.11.3 Potential for Replication in Other Regions

The success of saffron cultivation in Chitral suggests strong potential for replication in other mountainous areas in Gilgit-Baltistan, the highlands of Baluchistan, and even parts of northern Afghanistan. Lessons learned in community engagement, organic certification, and value-chain development can be applied in similar socio-ecological contexts.

8.12 Conclusion

8.12.1 The Saffron Pilot Project

The saffron cultivation pilot project in Upper Chitral demonstrates the viability of climate-smart regenerative agriculture in high-altitude settings challenged by limited cultivable land, water scarcity, and the looming threats of climate change. By systematically selecting suitable sites, engaging women growers, adhering to organic standards, and forging strategic partnerships for value addition, RegenX.eco has laid the groundwork for a sustainable saffron industry in Chitral.

From an environmental perspective, saffron's low water footprint and potential to enhance soil health through organic practices align closely with ecological conservation goals. Socially, the pilot's emphasis on women's involvement underscores the transformative power of specialty crops in empowering marginalized groups. Economically, the global saffron market provides significant economic opportunities for local farmers if they can differentiate their products via organic labeling, high-quality standards, and ethical sourcing narratives.

However, as with any nascent venture, scaling saffron production in Chitral will require addressing bottlenecks in infrastructure, policy support, and consistent bulb supply. Continued investment in research, training, and marketing investments will be essential to sustain momentum. With the robust foundation established by RegenX.eco and North Naturals, the stage is set for saffron to become a high value, climate-resilient alternative for communities across Chitral and beyond.

Ultimately, the success of saffron cultivation in Chitral underscores the broader potential of climate-smart regenerative agriculture in mountainous areas worldwide. By leveraging local knowledge, scientific innovation, and market linkages, once-marginalized regions can turn ecological vulnerabilities into economic strengths, fostering resilience in the face of accelerating climate change.

8.12.2 Contribution to Sustainable Regional Development

The saffron cultivation pilot project in Upper Chitral demonstrates how a broad-based, community-driven initiative can contribute to the sustainable development of a challenging mountain region. Key factors include community ownership of the initiative, professional support and monitoring, consideration of sociocultural aspects, and a focus on socioeconomic outcomes. Innovation, entrepreneurship, and a focus on value-creation systems will be essential for establishing a stable and sustainable economic future for the region.

References

Aung HH, Wang CZ, Ni M, Fishbein A, Mehendale SR, Xie JT, Shoyama CY, Yuan CS (2007) Crocin from Crocus sativus possesses significant anti-tumor effects against human colorectal cancer cells in vitro and in vivo. Cancer Lett 246(1):167–176. https://doi.org/10.1016/0304-3835(95)04067-6

Aziz Ali Dad (2016) Chitral community development network: a case study of the network of LSOs in Chitral. Retrieved 16 July 2025, from http://akrsp.org.pk/wp-content/uploads/2016/08/Case-Study-Chitral-Community-Development-Network.pdf

Bathaie SZ, Mousavi SZ (2010) New applications and mechanisms of action of saffron and its important ingredients. Crit Rev Food Sci Nutr 50(8):761–786. https://doi.org/10.1080/10408390902773003

Bolch T, Kulkarni A, Kääb A, Huggel C, Paul F, Cogley JG, Frey H, Kargel JS, Fujita K, Scheel M, Bajracharya S, Stoffel M (2012) The state and fate of Himalayan glaciers. Science 336(6079):310–314. https://doi.org/10.1126/science.1215828

CBD, Convention on Biological Diversity (2024) Topics. Retrieved 16 July 2025, from https://www.cbd.int/topics

FAO (2013) Climate-smart agriculture sourcebook. Retrieved 16 July 2025, from https://www.fao.org/climate-smart-agriculture-sourcebook/en/

FAO (2021) FAO statistical databases: production quantities of saffron. Food and Agriculture Organization of the United Nations. Retrieved 16 July 2025, from http://www.fao.org/faostat

Fernández J-A (2004) Biology, biotechnology and biomedicine of saffron. In: Recent research developments in plant science, vol 2, pp 127–159. https://doi.org/10.17660/actahortic.2004.650

Grand View Research (2020) Saffron market size, share & trends analysis report. Grand View Research

Hausenblas HA, Saha D, Dubyak PJ, Anton SD (2013) Saffron (Crocus sativus L.) and major depressive disorder: a meta-analysis of randomized clinical trials. J Integr Med 11(6):377–383. https://doi.org/10.3736/jintegrmed2013056

Kothari D, Thakur M, Joshi R, Kumar A, Kumar R (2021) Agro-climatic suitability evaluation for saffron production in areas of western Himalaya. Front Plant Sci 12:657819. https://doi.org/10.3389/fpls.2021.657819

Lutz AF, Immerzeel WW, Shrestha AB, Bierkens MF (2014) Consistent increase in high Asia's runoff due to increasing glacier melt and precipitation. Nat Clim Chang 4(7):587–592. https://doi.org/10.1038/nclimate2237

Melnyk JP, Wang S, Marcone MF (2010) Chemical and biological properties of the world's most expensive spice: saffron. Food Res Int 43(8):1981–1989. https://doi.org/10.1016/j.foodres.2010.07.033

Razavi BM, Hosseinzadeh H (2017) Saffron: a promising natural medicine in the treatment of metabolic syndrome. J Sci Food Agric 97(6):1679–1685. https://doi.org/10.1002/jsfa.8134
Sustainable Development (2024) Retrieved 16 July 2025, from https://sdgs.un.org/
Upper Chitral District (2024) Wikipedia. https://en.wikipedia.org/w/index.php?title=Upper_Chitral_District&oldid=1263416305
Wehrli A (2014) Why mountains matter for sustainable development. Mt Res Dev 34(4):405–409. https://doi.org/10.1659/MRD-JOURNAL-D-14-00096.1
Wymann von Dach S, Bracher CP, Peralvo M, Perez K, Adler C (2018) Leaving no one in mountains behind: localizing the SDGs for resilience of mountain people and ecosystems. pp 1–12. https://doi.org/10.7892/boris.120130

Dinar Wali Founder and Computational Researcher at RegenX.eco, Pakistan. He applies data science, artificial intelligence, and computational modeling to design and scale climate-smart, nature-based solutions. His work focuses on the use of geospatial analysis, climate data, and decision-support tools to advance regenerative agriculture, agroforestry, renewable energy, and related sustainability initiatives.

Peter Marty Lecturer and Senior Scientist at the Institute of Natural Resource Sciences at the Zurich University of Applied Sciences (ZHAW). His research and teaching focus on spatial planning, regional development, and sustainable development in rural and mountain regions. Drawing on extensive experience in landscape management, tourism, environmental planning, and regional policy, his work explores pathways for fostering resilient communities, sustainable land use, and balanced regional development in the face of social, economic, and environmental change.

Chapter 9
Hope Needs Courageous Action: Local-to-Local, Cross-Border Community-Academic Partnerships can Serve as Models for Social and Ecological Transformation

Iris T. Stewart

Abstract The concept of hope calls us to courageous action in the face of the largest global environmental crisis in human history. It is deeply embedded as a key element for positive change and renewal in Pope Francis' Encyclical *Laudato Si On care of our common home*. This article explores how, building on the Integral Ecology framework and environmental justice principles, community-academic collaborations across borders and community-based climate justice work can foster hope grounded in lived experiences in the context of intersecting ecological, social, and political crises. Drawing on spiritual and classical philosophical traditions, it argues that hope is not a passive state but a courageous practice rooted in action, agency, and solidarity. The Anthropocene is presented as an era of profound disruption to ecosystems and human well-being, highlighting the need for hopeful responses anchored in justice and interconnection. Particular attention is given to the work of the Water and Climate Justice Lab at Santa Clara University, which partners with smallholder communities in northern Nicaragua through long-term collaboration based on Community-Based Participatory Action Research (CBPAR). It details joint, local-to-local efforts to map local water systems, conduct climate vulnerability assessments, and document changing precipitation patterns in one of the driest and most climate-sensitive regions of Central America. The development of NicaAgua, a climate forecasting and early warning app co-designed with local partners, further demonstrates how complex climate data can be translated into accessible tools for adaptation. While such initiatives cannot solve global crises alone, they offer scalable models of hope in action for social and ecological transformation that are rooted in mutual respect, technological co-creation, and a shared commitment to flourishing human and ecological communities.

I. T. Stewart (✉)
Department of Environmental Studies and Sciences, Santa Clara University, Santa Clara, CA, USA
e-mail: istewartfrey@scu.edu

T. Karcher SJ et al. (eds.), *Making Hope Possible*, Environmental Humanities: Transformation, Governance, Ethics, Law 7,
https://doi.org/10.1007/978-3-032-17719-3_9

Keywords Environmental justice · Community-engaged research · Sustainability · Hope · Climate adaptation

9.1 Hope as a Courageous Practice

What does it mean to *create hope together* in a time of planetary crisis?

While humans have dealt with catastrophes of biblical proportions throughout their history—floods, plague, and strife, to name a few—as a human community, we now find ourselves in the Anthropocene. And at the current moment, living in the Anthropocene is not good news. It describes a newly designated epoch in Earth's history, in which the human footprint is documented in the Earth's archives. This footprint, including toxic soil layers, contamination recorded in ice and sediments, and a significantly altered atmosphere and climate, is diminishing a broad range of ecosystem services that are essential for human and non-human life on Earth (Steffen et al. 2011). Essential ecosystem services are founded on a web of life and physical processes that are connected and depend on each other. For example, in California mountain streams, the native salmon that once were abundant need clean and cold water, shaded spawning and rearing grounds, small insects for food, and the ability to travel upstream. In return, they transport nutrients, fertility, and food from the ocean into the upper reaches of the watershed. Many salmon runs have been wiped out or reduced from several hundreds of thousands to a couple of thousand. Similar stories abound from all over the globe. With ecosystems and their keystone species vanishing, their services, including the provision of food and water (provisioning services), regulating climate and air quality (regulating services), maintenance of healthy soil and pollination (supporting services), and offering of recreational and cultural benefits (cultural services) are vanishing too. All of these are crucial for human survival, but also more than that, they are crucial for human well-being.

The reality of the Anthropocene has only been formally acknowledged for a couple of decades, but its origins are widely accepted to lie in the Industrial revolution, and the explosion of carbon emissions, toxins, and forever chemicals such as DDT, that are now found in the most remote corners of the planet such as Antarctica (Geisz et al. 2008). At the same time, we are sharing the bounties of this planet, such as clean water, clean air, beneficial rains and temperatures, oceans, mountains, and river valleys with an increasing number of human family members (Prior et al. 2012). The capacity of man to modify the physical environment where he lives has exponentially grown with technological advances. While many of these technological advances can support the flourishing of the human species, and, arguably, those of ecosystems too under the right conditions, a concurrent capacity in humans for a greedy extractivism has surpassed any kind of limitation. As a result, for many people the hope for a better future has shifted to expectations of societal and ecological catastrophes, with some hope of holding on to the status quo as long as possible. Holding on to the status quo, though, includes holding on to the very uneven environmental benefits and burdens that concepts such as environmental justice and Laudato Si seek to address.

Interestingly, the philosophers of ancient Greece already recognized two different terms for hope that reflect this very modern shift in our expectations for the future. One referred to the expectation of an outcome to take place in response to our actions, whether this outcome be positive or negative ("Elpis" or anticipation), and a second, slightly adjusted term designating a positive outcome ("Eulpis," or positive anticipation, where "eu" is rooted in "good."). The first one is what we as humankind appear to be practicing as a collective mantra: We understand the impact of our actions when it comes to pollution, greed, war, inequity, destruction, and extinction, and yet we continue to let ourselves be swept into that direction. The second term for "positive hope" in classical Greek recognizes the importance placed on human agency, deliberation, and courage as a critical component of hope, and hopefulness as a general foundation for a good life (Gravlee 2020). Hope arises in situations when we do not know what will happen, it is invariably connected to the future, yet hope in its classical grounding is not a vague and unfounded notion fed by wishful thinking and ignorance, it has a moral rationality and justifiable aims that possess the ability to sustain and motivate us.

As Aristotle notes, "positive hope" calls us to courageous action:

> *'A coward is a pessimistic sort of fellow, for he fears everything. But a courageous man is the very opposite, because confidence implies hopefulness'*. —Aristotle, NE 3.7, 1116a3

Since the earliest Christian writers, hope has been considered as one of the three major Christian virtues. Hope and hopefulness have several contexts, but the most important of these is courage, and courage for action. In Pope Francis' encyclical *Laudato Si'* (2015), arguably the most important text on Catholic social teaching and the current, rapidly accelerating global environmental crisis, the concept of hope through action is deeply embedded. He in particular emphasizes that despite the largest global environmental crisis in human history, there is still potential for positive change and renewal. Francis notes that

> '*Hope would have us recognize that there is always a way out, that we can always redirect our steps, that we can always do something to solve our problems*' (61) and '*The Creator does not abandon us; he never forsakes his loving plan or repents of having created us. Humanity still has the ability to work together in building our common home* (13).'
>
> —Pope Francis, *Laudato Si'*—On care for our common home.

In the Pope's teaching, hope springs from the faith for a better world through community, action, and trust. It asks us to start by who we are and where we are. *Laudato Si'* introduces a new framework, termed "Integral Ecology," which recognizes that what is needed is an integration of spiritual, ecological, and social transformation to address crises from the local to the global scale. Jesuit institutions, like Santa Clara University, are guided by Jesuit values of dignity for all human beings, solidarity, participation, compassion, and justice which we may bring to our work and to creating a hope-filled future. The work of the Environmental Justice and the Common Good Initiative, and its affiliate, the Water and Climate Justice lab, seek to co-create cross-border partnerships that put these values into practice and could serve as models for building a world where both humans and ecosystems thrive.

9.2 Creating Pathways for Hope in a Time of Planetary Crisis

Inspired by *Laudato Si'*, and principles of both Integral Ecology and environmental justice, the Environmental Justice and the Common Good Initiative (EJ&CGI) at Santa Clara provides community-driven research and learning for environmental and social justice. As an affiliate of the EJ&CGI, the Water and Climate Justice lab (WCJ) at Santa Clara University follows the principles of CBPAR (Community-Based Participatory Action Research) with the aim of furthering collaboration and community-academic partnerships for water and climate justice through multidisciplinary and student-centered collaborations. The CBPAR approach is centered on creating community partnerships such that community representatives participate in the definition of the research problem, the selection and carrying out of methods, interpretation of the data, and application of the findings may help address these concerns. For projects, the EJ&CGI and the WCJ lab also prioritize obtaining funding for the work of our community partners, and support their grant applications (Fig. 9.1).

One of the focal areas of the work of the WCJ lab is in Northern Nicaragua, where a 10-year collaboration with a local social development organization, CII-ASDENIC. CII-ASDENIC works with smallholder farming communities in the Segovias region of Nicaragua has been developed.

Fig. 9.1 Practical application of community-based participatory action research. Figure co-developed by Iris T. Stewart and Allan Baez Morales, implemented by Arturo Torres Torres Landa

As a country, Nicaragua has contributed very little to environmental pollution, habitat destruction and species extinction, and the emissions of greenhouse gases. Yet, it is extremely vulnerable to its effects. Nicaragua lies on the Central American isthmus and from a climate perspective is influenced by both the Atlantic and the Pacific oceanic systems, as well as the seasonal movement of the Inter-Tropical Convergence Zone (Stewart et al. 2021). It also lies in the pathway of tropical storms that have been more frequent and intense with warming oceans and its shores are affected by sea level rise and saltwater intrusion (Hidalgo et al. 2023). While in other parts of the world, ecosystems and coffee farmers have adapted to warmer temperatures by moving to higher elevations, the topography of Nicaragua limits that possibility. Volcanic areas of low mountains are mostly present in the north-western corner of the country.

At the same time, Nicaragua has had a long history of political conflict, and the current Ortega regime is re-shaping society in ways that quell democracy and violate human rights and freedom of expression (Golob 2024). This development has accelerated over the past 3 years, curtailing social and economic development, and impeding critical work toward alleviating poverty, hunger, barriers to access of safe water, and adaptation to climate change. It has also mostly limited our ability for in-person collaborations and exchanges, which were possible at the beginning of our engagement. We have needed to rethink how we can bridge borders and sustain and grow our partnership, what strategies toward sustainable social-ecological systems and technologies might be developed not in spite of, but because of the conditions we are finding ourselves in, and how they might be serving in other situations.

The communities we work with through CII-ASDENIC consist of 800–1500 people that produce coffee mainly for export, and beans, maize, fruits and vegetables, and small amounts of meat for local markets and their own consumption. The fields and gardens are almost exclusively rainfed, making the agricultural calendar dependent on the timing and the amount of precipitation. The communities in question are located around a more elevated forested plateau, which is protected as a nature reserve and acts as the local "water tower," capturing a larger share of the precipitation than lower lying areas. Small water systems tap the springs at the side of the plateau and divert this water to concrete storage tanks. Tanks are usually filled overnight and taps opened in the morning, when gravity drains them to spigots for community taps and individual households. The water is mostly, but not always, treated with chlorine by designated members of the community that manage the tank.

Beginning in 2014, the Water and Climate Justice team, together with CII-ASDENIC have carried out two surveys and held a series of workshops and focus groups in the communities to learn about community concerns and needs. Periodically, we have also shared results from an analysis of climate impacts we have conducted for Central America. The first of these surveys focused on understanding water access and water governance through water committees. The second survey sought to understand how communities are impacted by extreme weather and climate events, and identify current mitigation strategies and needs.

One of the requests from the communities was to have a mapped layout of their water sources, such as springs, wells, water holes, and streams, and their water systems. These mapped layouts were a useful tool to better understand how water resources may be shared from the upstream to the downstream and also sustainably managed. Community members and especially youth were involved in collecting GPS data points, which students at Santa Clara University then helped to convert into GIS layers and archive. Nicaraguan youth were empowered through the use of GPS sensors, technology, and the purpose for their communities; Santa Clara University students learned with empathic reflection about the challenges of water access in the global south. This is the first time the communities were able to access these kind of layouts to talk about shared management (Fig. 9.2).

Results from our water governance system survey indicated that each community has water committees that regulate the functioning, maintenance, access, and water needs of the community. When women were part of the committee, a greater sense of equity in access to water was reported. Together with community members, we also thought through existing and potential governing norms around water pollution and water sharing. For example, animal manure or household or agricultural chemicals may pollute the water downstream, and pumping water out of streams upstream leads to greater water insecurity in the communities downstream.

Relevant climatic metrics, recent climatic shifts, and their impact on household water security in Central America are not well understood due to sparse observations that impede both the analysis of these shifts and the impacts on the ground.

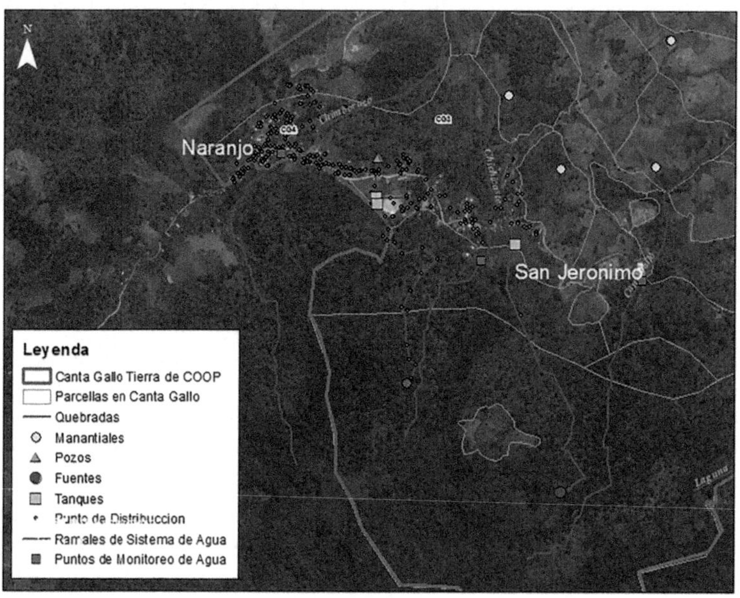

Fig. 9.2 Example of mapping community water sources and water systems. As results are shared with communities, maps and other products are created in Spanish (created by Iris T. Stewart)

Based on focus groups facilitated by CII-ASDENIC, we learned that two climate phenomena are important metrics: The climatic patterns of the Central American Dry Corridor (CADC), and the presence, timing, and intensity of the Mid-Summer Drought (MSD)—a lull in the precipitation during the wet summer months, that in a good year separates two rainy seasons that allow for two crops. We used local definitions of the CADC and the MSD to, for the first time, map and quantify climatic changes within the CADC (Stewart et al. 2021) and with the MSD (Maurer et al. 2022). Results were shared back to the communities and have formed the basis for a dialog about strategies for climate change adaptation and mitigation (Fig. 9.3).

Our analysis of the climate and survey data indicates that our partner communities in northern Nicaragua are in one of the drier regions of the CADC, an already water-limited region, with only seasonal rainfall, and are particularly vulnerable to further drying. They are one of the places in the Global South, where climate adaptation measures may be most urgent. Our work documents trends toward summer drying, increases in drought and aridity driven by precipitation declines, and a lengthening of the winter dry season. We then compare the experiences in water security and adaptation measures to climatic shifts for different communities. The primary water sources for drinking, cooking, bathing, and cleaning in the region are springs that feed small water systems, small streams, at times shared between communities. From both of our surveys and multiple focus groups we learned that experiences of water insecurity are common toward the end of the dry season (predominantly Feb–May) and are exacerbated during drought periods. Households employ diversified adaptation strategies to supplement or replace primary water sources during extended times of scarcity.

Through our second survey and in multiple focus groups, smallholder farming communities expressed concern about the impacts of climatic changes, particularly shifts in precipitation timing, longer and drier dry periods, and more frequent extreme

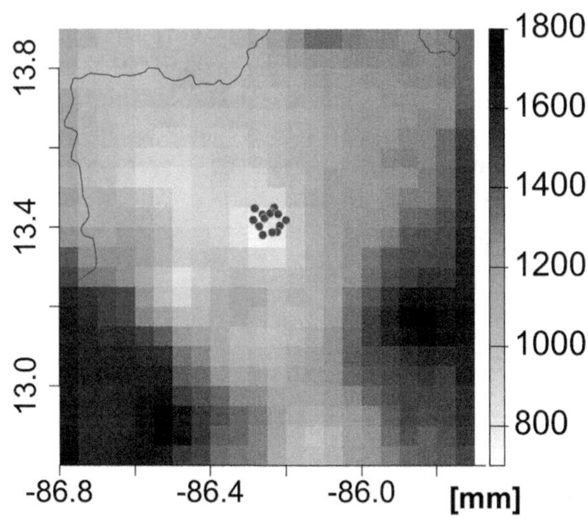

Fig. 9.3 Average annual precipitation and location of the communities within some of the driest area of the Central American Dry Corridor. Figure co-developed by Iris T. Stewart and Briana Guingona

rains and droughts, that were identified from climate projections. Repeatedly, they requested better information and climate forecasts as a basis for strengthening their climate resilience.

Three years ago, we partnered with the Frugal Innovation Hub at Santa Clara University to start the development of a weather and climate forecasting and early warning app for Nicaragua. The name of the app is NicaAgua', a fusion of "Nicaragua" and "Agua" which means water in Spanish. The Frugal Innovation Hub had explored an app prototype for local farmers that was bringing the metrics from a local weather station online. With ongoing collaboration by CII-ASDENIC, our work converged with the idea of providing short- and seasonal climate forecasts, as well as alerts when forecasts were outside the historic normal range, the capacity of CII-ASDENIC to send warning to communities, comparisons to the climate of the past 40 years, weather information from a local weather station. At every step of the way, and following the principles of CBPAR, these ideas were brought back and were discussed and tried at community workshops that CII-ASDENIC organized (Fig. 9.4).

The app is now currently online for Android smartphones and has been completely overhauled, thanks to a transdisciplinary group of four faculty and over 20 students from computer engineering, environmental studies and sciences, civil, environmental, and sustainable engineering, and web design. A user survey allowed us to have a better understanding of the people it is serving. For example, we identified that all respondents had at least one smartphone at home, and reported constant (17%), almost constant (28%), or at least intermittent (50%) internet access. While female and also older respondents indicated lower cell phone and internet access, younger respondents (<40 years of age) reported wide familiarity with the installation and use of apps and online resources (91%). CII-Asdenic, with our support continues to hold workshops in various communities of the region to introduce the app and facilitate conversations on climate resilience. Our next steps include pathways to translate weather and climate forecasts into agricultural responses. A first follow-up request by the farmers was to include the phase of the moon in the forecast to support agricultural decision-making based on more traditional knowledge. We have implemented that request and are looking into co-developing and implementing recommendations for specific preparations and responses to climatic hazards, such as water capture and food storage to prepare for drought, reinforcements against storms, adjustments to agricultural practices to adjust to shifts in temperature and the timing and precipitation.

9.3 Discussion

While the political developments toward authoritarian rule in Nicaragua are particularly concerning, and the country's struggle for economic and social development continues to grow, similar processes in both the Global South and North are pervasive, and have intensified since the publication of Laudato Si′ (Francis 2015). Besides a

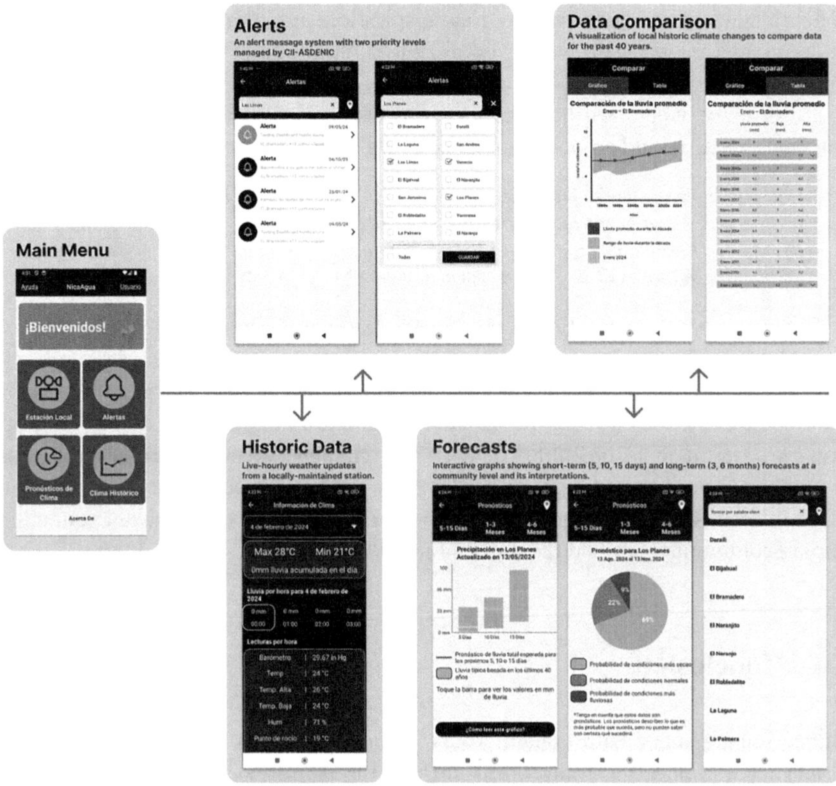

Fig. 9.4 Functionalities of the "NicaAgua" climate forecasting, climate change analysis, and early warning app for communities in northern Nicaragua. Figure co-developed by Allan Baez Morales, Arturo Torres Torres Landa, and Iris T. Stewart

global ecological crisis, and a global failure of reaching the millennium development goals for a majority of the world's population, the strength of democratic governments is sliding internally and abroad. What becomes evident is a growing need for social development that has been connected to happiness and hope (Veenhofen 2024), and is represented by stable institutions, an independent and fair judiciary, the freedom of political and faith expression, and policies that actively prioritize improving the well-being of its citizens by focusing on aspects like education, healthcare, social equality, community development, and overall quality of life, aiming to create a more inclusive and equitable society.

To truly hope in the face of our current ecological and humanitarian crises takes courage in the face of real concern, a rejection of despair, an acknowledgment of our own weakness and vulnerability, and the choice to move in the direction of a positive outcome anyway. Community-based collaborations across borders and lived experiences may provide pathways toward these positive outcomes on a small scale. Are they enough? No, we cannot be kidding ourselves. But they could serve as models and strategies from which we can draw inspiration and faith, from which we might scale up, and from which we can connect to a multitude of other nodes of hope to find a web.

In terms of where we have gone wrong, Francis (2015) points to our loss of spiritual guidance and an overabundance of possibility.

> *'Both are ultimately due to the same evil: the notion that there are no indisputable truths to guide our lives, and hence human freedom is limitless. We have forgotten that "man is not only a freedom which he creates for himself. Man does not create himself. He is spirit and will, but also nature"* (12). —Pope Francis, Laudato Si'—On care for our common home.

If the source of destruction is the limitless human freedom to destruct, then we might address this ill with a nature-based homeopathic approach: Sharing some well-guided doses of this freedom, not as a freedom to dominate, take, and destroy, but the freedom of choosing well-being, service, community, human survival. As Pope Francis states, man is constituted of spirit and will and also our nature. If we are able to go back to our true nature, we find our place within the ecological web, not above it, we find our spirituality and faith, we find the critical need for survival in our entire human community, we find the quiet, hopeful voice within.

9.4 Conclusions

Human engagement with contemporary planetary distress is informed by multi-dimensional conceptual frameworks such as the Anthropocene, philosophical and theological interpretations of hope as a moral imperative, and the tangible socio-ecological challenges unfolding worldwide, particularly in the Global South. What emerges is a recognition that the current sweeping environmental and societal crises are intersecting and accelerating, that they require courageous actions motivated by hope on local to global scales, and that a transformation toward flourishing social and ecological systems must be based on care, justice, and mutual respect. Pope Francis' encyclical, *Laudato Si'*, which has integrated social, ecological, and spiritual transformation into a new framework, Integral Ecology, provides a new philosophical dimension in this conversation. While *Laudato Si'* is rooted in the Christian tradition, it is also a universal invitation and call to action.

In this context, the case study from Nicaragua presented here aims to represent more than a study in applied environmental justice, it seeks to integrate the theoretical with developing pathways to manifesting Integral Ecology in action. The work has sought to respond to planetary crises not with despair, but with a deliberate practice of "Eulpis"—a rational, morally grounded hope that emerges from community, agency, and co-creation. Through its long-term collaboration with CII-ASDENIC and the application of CBPAR principles, the Water and Climate Justice Lab has fostered a process where knowledge is not merely extracted, but exchanged and owned. Through such projects, hope can materialize from the ground up in tangible ways: in women's active roles in water governance, in the fusion of traditional ecological knowledge with digital tools, and in the moral imperative to treat the Earth and one another with care and dignity. These efforts align fully with the call of *Laudato Si′*, emphasizing that ecological, social, and spiritual renewal must happen together. What makes this case study worthy of consideration is not that it solves a part of the global crisis, but that it demonstrates how even under conditions of political repression and ecological vulnerability, hope can be cultivated through trust, mutual learning, and the commitment to showing up. This is not abstract optimism, it is a model for courageous action that might serve as a node in a broader web of global transformation toward thriving human and ecological communities. If we are to imagine a future worth living, it will be built not only through policy and protest, but through partnerships like this, where hope is not only envisioned, but enacted, together.

Acknowledgments I am deeply grateful to my colleagues in the Water and Climate Justice lab, particularly Allan Baez Morales, Jake Dialesandro, and Ed Maurer for their collaborations on community engagement or app development. Many important contributions to this work have been made by that, especially Briana Guingona, Gautam Chitnis, and Arturo Torres Torres Landa. I have learned much from my colleagues at the Environmental Justice and the Common Good Initiative, Chris Bacon, Chad Raphael, Jesica Fernandez and Rocio Segura on community-engaged environmental justice research. Funding for this work by the Whitham Foundation, the Miller Center for Global Engagement, and the Environmental Justice and the Common Good Initiative at Santa Clara University is gratefully acknowledged.

References

Francis P (2015) Laudato Si' [on Care for our Common Home] [encyclical letter]. http://www.vatican.va/content/francesco/en/encyclicals/documents/papa-francesco_20150524_enciclica-laudato-si.html

Geisz HN, Dickhut RM, Cochran MA, Fraser WR, Ducklow HW (2008) Melting glaciers: a probable source of DDT to the Antarctic marine ecosystem. Environ Sci Technol 42(11):3958–3962. https://doi.org/10.1021/es702919n

Golob A (2024) Nicaragua's new dictatorship: impacts of authoritarian rule on the health sector and civil society. Soc Sci Humanit Open. https://doi.org/10.1016/j.ssaho.2024.100878

Gravlee GS (2020) Hope in ancient Greek philosophy. In: Historical and multidisciplinary perspectives on hope, pp 3–23. https://doi.org/10.1007/978-3-030-46489-9_1

Hidalgo HG, Alfaro EJ, Valverde KT, Bazo J (2023) Probability of induced extreme precipitation events in Central America due to tropical cyclone positions in the surrounding oceans. Nat Hazards 116(3):2917–2933. https://doi.org/10.1007/s11069-022-05790-1

Maurer EP, Stewart IT, Joseph K, Hidalgo HG (2022) The Mesoamerican mid-summer drought: the impact of its definition on occurrences and recent changes. Hydrol Earth Syst Sci 26(5):1425–1437. https://doi.org/10.5194/hess-26-1425-2022

Prior T, Giurco D, Mudd G, Mason L, Behrisch J (2012) Resource depletion, peak minerals and the implications for sustainable resource management. Glob Environ Change 22(3):577–558. https://doi.org/10.1016/j.gloenvcha.2011.08.009

Steffen W, Grinevald J, Crutzen P, McNeill J (2011) The Anthropocene: conceptual and historical perspectives. Philos Trans R Soc A Math Phys Eng Sci 369(1938):842–867. https://doi.org/10.1098/rsta.2010.0327

Stewart IT, Maurer EP, Stahl K, Joseph K (2021) Recent evidence for warmer and drier growing seasons in climate sensitive regions of Central America from multiple global datasets. Int J Climatol 42(3):1399–1417. https://doi.org/10.1002/joc.7310

Veenhoven R (2024) World database of happiness. In: Encyclopedia of quality of life and Well-being research. Springer, Cham, pp 7861–7865. https://doi.org/10.1007/978-94-007-0753-5

Iris T. Stewart Professor in the Department Environmental Studies & Sciences at Santa Clara University, USA. Her research focuses on the impacts of pollution, climate variability and climate change on water resources, ecosystems, and human communities. Combining hydrology, spatial analysis, and community-based research, she explores questions of water security, environmental justice, and climate resilience, with particular attention to the equitable distribution of environmental benefits and burdens and the development of sustainable water and climate systems.

Chapter 10
Everything is Interconnected: Understanding Investor-State Dispute Settlement as an Impediment to an Integral Ecology in Africa

Fernando C. Saldivar, S.J.

Abstract This article examines Investor-State Dispute Settlement (ISDS) through the lens of integral ecology, emphasizing its impact on climate justice in African and Global South states. The introduction outlines how ISDS challenges sovereign climate action. The Catholic Church's critique of the global financial order in the era of Pope Francis, grounded in integral ecology principles, provides a framework for understanding these challenges. A detailed overview of ISDS follows, explaining its legal basis, the relationship between trade, foreign direct investment and international investment treaties, and the arbitration process. This section highlights how ISDS empowers private investors to challenge governmental climate measures, often resulting in significant financial liabilities that limit regulatory sovereignty, creating a chilling effect on climate legislation in Africa. Finally, the discussion proposes hope as a bridge between legal structures and theological ethics, pointing toward an integral ecology approach that aligns law with environmental stewardship. The article emphasizes the need for greater awareness among climate activists of ISDS's structural constraints on environmental justice.

Keywords Investor-state dispute settlement · Integral ecology · Africa · International investment law · Global South · Climate · Foreign direct investment · Catholic social teaching · Pope Francis

---

Where he is affiliated with the international human rights center

---

F. C. Saldivar, S.J. (✉)
LMU Loyola Law School, Los Angeles, CA, USA
e-mail: fsaldivar@jesuits.org

T. Karcher SJ et al. (eds.), *Making Hope Possible*, Environmental Humanities: Transformation, Governance, Ethics, Law 7,
https://doi.org/10.1007/978-3-032-17719-3_10

10.1 Introduction

In his 2015 encyclical on Care for our Common Home *Laudato Si'*, Pope Francis develops *integral ecology* as his fundamental understanding that *"everything is closely interrelated"* and that *"today's problems call for a vision capable of taking into account every aspect of the global crisis"* (Francis 2015, sec. 137). Although care for creation is its beating heart, integral ecology is not limited to the environment. If everything is truly connected and interrelated in all aspects, then the crises we face are not separate. We should not think in terms of "one environmental and the other social, but rather one complex crisis which is both social and environmental" wherein the policy solutions to address climate change, poverty, and development will need to be multidimensional and interdisciplinary (Francis 2015, sec. 139).

Catholic social teaching, in general, and the magisterial documents issued by popes over the last century and a half, of which encyclicals like *Laudato Si'* have the greatest normative value within the Church, are an underappreciated point of departure for transdisciplinary critique of the normative structures of international politics and finance. This is because popes going back to Leo XII's 1891 encyclical *Rerum Novarum* (on capital and labor), through the two world wars and up to the present, have issued these documents and elaborated a Catholic social teaching to speak directly to the world at large on critical issues of the age. These documents specifically intended to inform the reflection of global policymakers by emphasizing fundamental human dignity, the common good, and solidarity (de Charentenay 2008). In the era of Pope Francis, a specific focus of this teaching has been a critique of the global financial system, resulting in the publication of *Oeconomicae et pecunaiarie quaestiones* in 2018 by the Vatican's Congregation for the Doctrine of the Faith and the Dicastery for the Promotion of Integral Human Development, a document offering a reflection on certain aspects of current economic-financial systems (*Oeconomicaę et pecuniariae quaestiones* 2018). Notwithstanding, an integral ecology based reflection on international law, specifically as it relates to the normative structure underpinning the global financial system, has yet to fully develop.

This is not to say that either *Laudato Si'*, integral ecology, or any other aspect of Catholic social teaching on the economy is absent from legal scholarship in the United States (Flournoy 2020). Rather, to the extent it is engaged at all, it is done in a way that somehow secularizes it, treating Catholic social teaching simply as any other policy proposal among many others, thus minimizing these teachings' import as works of Christian theological ethics (Ramirez 2017). Thus, the moral and ethical dimension which underpins these documents is trivialized, which is a shame given the privileged space that Catholic social teaching operates in and through which it is able to bridge the gap between its "natural terrain" of theology, scripture, and dogma and the secular concerns of society, politics, and economics (de Charentenay 2008, p. 7).

Law gets somewhat squeamish in the neighborhood of God-talk (Cook 1994). However, that is the nature of what Pope Francis means by an *integral* ecology. Integral must include the spiritual. "Is it reasonable and enlightened," Francis asks, "to dismiss certain writings simply because they arose in the context of religious belief?" (Francis 2015, sec. 199). This, perhaps, is what makes integral ecology such a radical way of reconceptualizing the global economic and financial order. The climate crisis indeed itself provides a liminal space where law and theology have something profound and timely to say to one another. Nevertheless, it is important to note that integral ecology is more than simply the spiritualization or addition of a religious element to ecology, a point which Pope Francis emphasized in *Laudato Si'*. Rather, it is an understanding that acknowledging the spiritual dimension of the ecological crisis, in all of its forms, better informs our reflection on its other aspects.

As it stands, a decade after the publication of *Laudato Si'*, the legal academy, specifically as it relates to business and human rights discourse, has yet to be fully engaged with theological ethics as a conversation partner for incarnating Francis's vision of integral ecology. Serious work remains to be done in using integral ecology as a framework for critiquing the global structural injustices inherent in the relationship between states, corporations, business and finance, particularly as it applies to Africa and the Global South broadly speaking. This is a critical gap that needs to be filled, and the resulting dialog can help to bolster a wavering commitment around the world to the promotion of human dignity and the common good, two of the pillars on which integral ecology rests, and which Anna Rowlands suggests can help us build a politics of communion (Rowlands 2021).

This article will undertake just such an interdisciplinary analysis, exploring the use of Investor-State Dispute Settlement (ISDS) through an integral ecology lens. ISDS is an area of international investment law that has a tremendous impact on the ability of African states, and those in the Global South generally, to mitigate and adapt to climate change in that it provides a private, nontransparent arbitral forum for *individual* foreign investors to challenge measures undertaken by governments which negatively affect their profit expectations, often resulting in crippling monetary damages against the host state. ISDS is not well understood by climate stakeholders outside the legal and business communities but is particularly problematic on a structural level in that it gives private individuals and foreign corporations an effective veto over climate legislation in developing countries. This results in a regulatory chilling effect which calls for a greater awareness by the activist community.

Whatever one may think of the health of the current rules-based global order, it exists and its rules matter. This article will provide to an audience invested in integral ecology principles an example of how efforts to bring this vision to life will be futile unless our perspective is widened to look at key structural deficiencies and normative injustices in the global financial and political order.

10.2 Integral Ecology and the Catholic Church's Critique of the Global Financial Order

It is no exaggeration to say that *Laudato Si'* marks something truly revolutionary in the Catholic Church's modern social teaching: the first papal encyclical dedicated "explicitly and entirely to the ecological question" (Euvé 2021, p. 65). However, it is not simply an "ecological" or "green" encyclical, but is firmly rooted in the entirety of that social teaching, the modern period of which opens up with Pope Leo XII's encyclical on the situation of labor, *Rerum Novarum* in 1891 (Euvé 2021, p. 145). In fact, this "gathering momentum" for the Church's "gradually developing body of CST [Catholic social teaching] on ecological issues" has been taking shape since the papacy of John Paul II (Rowlands 2021, p. 270).

Particularly, the mold around which this teaching takes shape is John Paul II's 1987 encyclical dedicated to the issue of development, *Sollicitudo rei socialis*, where he "uses solidarity as a major concept to elaborate his vision of full development and the challenges of the late twentieth-century world" (Catta 2019, p. 45). There, Pope John Paul II develops his idea of "structures of sin" which he insists always need to be considered in relation to personal sin. These structures of sin "are rooted in personal sin, and thus always linked to the concrete acts of individuals who introduce these structures, consolidate them and make them difficult to remove" (John Paul II 1987, sec. 36). However, John Paul II makes a key point that will have an impact on how integral ecology develops under Francis, specifically that the personal sin does not simply lie in some distant past that produced the actual structures, but also "[n]ot acting against them or remaining voluntarily blind to them is also sinful" (Catta 2019, p. 62). According to John Paul II, "…just as one may sin through selfishness and the desire for excessive profit and power, one may also be found wanting with regard to the urgent needs of multitudes of human beings submerged in conditions of underdevelopment, through fear, indecision, and, basically, through cowardice" (John Paul II 1987, sec. 36). Failure to act in the face of injustice is no less sinful than deliberate action.

Nevertheless, there is no mistaking the fact that *Laudato Si'* is something different than what came before. In choosing "Francis," in reference to Francis of Assisi, as his papal name, Grégoire Catta says that Jorge Bergoglio gave us some indication of how his attention as pope would be focused. "Attention to the poor, to creation, and to peace cannot be separated and ought to drive the changes necessary in the world" (Catta 2019, p. 116). This rich Franciscan veneer, however, is superimposed upon something else that separates this pope from all of his predecessors—the fact that he is a Jesuit and from Latin America. Both of these shape his ecclesiology, his ideas about governance, and, perhaps most importantly, his manner of prayer (Deck 2022). In ways large and small, the Spiritual Exercises of Ignatius of Loyola shape his decision-making process and his manner of reflection, a lens which gives us a perspective on how to read *Laudato Si'* (Bernardi 2017). Nowhere does this become more clear than with trying to decipher what Francis is trying to express in his development of integral ecology (Deck 2022).

That is because in *Laudato Si'* Francis gives us more of a framework for *discernment*, a vision or a dream of a different future together, rather than a set of regulations to implement (Francis and Ivereigh 2020). Thus, in framing his integral ecology by saying, "It cannot be emphasized enough how everything is interconnected," Francis is giving us a prism for thinking about *ecology*, which he defines as "the relationship between living organisms and the environment in which they develop [which] necessarily entails reflection and debate about the conditions required for the life and survival of society, and the honesty needed to question certain models of development, production, and consumption" (Francis 2015, sec. 138). The interconnectedness of everything is the thread which connects our triple critique of development, production, and consumption, while also presenting challenges for us.

According to Pope Francis, "a true ecological approach *always* becomes a social approach; it must integrate questions of justice in debates on the environment, so as to hear *both the cry of the earth and the cry of the poor*" (Francis 2015, sec. 49). In fact, at the outset of the encyclical, Francis notes that "inequity affects not only individuals but entire countries" compelling us to discern an ethics of international relations that considers the "ecological debt" that exists between the global north and south (Francis 2015, sec. 51). Cardinal Michael Czerny and Christian Barone argue that Francis's social teaching provides five criteria for the worldwide Church and "each particular Church to undertake a resolute process of discernment, purification, and reform" (Czerny and Barone 2022, p. 45). The fifth of these criteria is "care for creation," which itself rests on the premise that "[t]he socio-environmental sphere struggles to overcome the ills of an economic-financial system that generates global inequalities" (Czerny and Barone 2022, p. 60).

In fact, Francis's critique of global economic structures is so sharp that it would not be unfair to characterize *Laudato Si'* as "anti-systemic" to the extent that it specifically targets the structures of financial capitalism as those that aid and abet these inequalities and injustices (Lobez 2022, p. 129). However, the Pope's purpose is not simply to throw stones at bankers or the financial industry—a point he has repeatedly emphasized. Instead, he recognizes that "[p]olitics must not be subject to the economy, nor should the economy be subject to the dictates of an efficiency-driven paradigm of technocracy…[Instead] there is an urgent need for politics and economics to enter into a frank dialog in the service of life, especially human life" (Francis 2015, sec. 189).

What, then, is the fruit of this dialog between politics and the economy from an integral ecology perspective? It is the recognition, quite simply, that *everything is interconnected*, that decisions made in the corridors of power in the Global North can, and do, have a significant and detrimental impact on individuals and communities in the South. It is the fact that the problems and concerns of the majority of the world's population, billions of people outside of Europe and North America, become the "collateral damage" of those very same decisions (Francis 2015, sec. 49).

The problem, as the Pope sees it, is structural. It is the "lack of a culture needed to confront this crisis" (Francis 2015, sec. 53). Thus, "[t]he establishment of a legal framework which can set clear boundaries and ensure the protection of ecosystems has become indispensable; otherwise, the new power structures based on the technocratic paradigm may overwhelm not only our politics, but also freedom and justice" (Francis 2015, sec. 53). "If everything is related, then the health of a society's institutions has consequences for the environment and the quality of human life" (Francis 2015, sec. 142). While building a culture of respect for the law, particularly as it applies to or implicates environmental regulation, is important, equally critical is a review of the *existing* normative framework underlying the international economic and financial order from an integral ecology perspective. Such a critique is desperately needed, because while Francis's vision of an integral ecology does call for individual ecological conversion, only to the extent that that leads to *structural* conversion will there be the changes necessary to mitigate and adapt to the cataclysmic effects of climate change worldwide (Renouard and de Bénazé 2023).

10.3 What is Investor-State Dispute Settlement (ISDS) and Why Does It Matter for Environmental Justice?

10.3.1 Trade Flows and the World Economy

Despite political rhetoric in many countries touting national resilience and deglobalization, especially in the wake of the COVID-19 global pandemic, economic data bears out the exact opposite—the world economy is becoming *more* interconnected, not less (Setser 2024). In fact, COVID-19 and the economic shutdown it generated only revealed the extent to which the global economy is radically interconnected and the extent to which decisions made in one state carry over into others (Tooze 2021). According to the World Trade Organization (WTO), a "Great Convergence" has taken place in the global economy which "was only possible because the world economy has grown more open and integrated, driven mainly by declining transport and communications costs, but also by decreasing trade barriers and increasing economic cooperation, all underpinned by a rules-based international economic system (WTO 2024, p. 21).

While the inequalities produced and sustained by this "Great Convergence" are themselves subject to critique and reform, there is one particular aspect of the WTO's statement that should draw our attention here: the fact that it is underpinned by a *rules-based* international economic system. The rules matter, as well as a myriad of questions that flow from that. Namely: Who makes the rules? What are they? Who interprets them? How are they enforced? How do you change them? In the most powerful economic states, particularly the United States, the line between economic and foreign policy is increasingly blurred, especially in the years since the 2008 financial crisis (Tooze 2019).

Whatever one may think of the rules-based economic and financial order, it *exists* and its rules matter. Thus, a structural level critique of the global economy such as Pope Francis calls for in *Laudato Si'* requires an understanding of the normative scaffolding on which it is constructed. In this case, that calls for a closer look at international trade law—specifically how disputes are adjudicated between foreign investors and the states being invested in.

10.3.2 The Nexus Between Trade, Foreign Direct Investment (FDI), and International Investment Treaties (IITs)

Although we customarily think of international trade as the exchange of goods and services across national borders, there is more to the system than the fleets of cargo ships traversing the world's seas at any particular moment in time. In a sense, there is the level in the supply chain which comes above these tangible end-products. That level is represented by capital flows. This phenomenon is the movement of money and investments across international borders, which permits all of this production to take place: the *financial* aspect of global trade (Obstfeld and Taylor 2017).

Foreign Direct Investment (FDI) is the engine for much of this globalization of trade (WTO 2024). FDI is the process where an investor in one country (an individual or a corporation) establishes a lasting interest and a significant degree of influence over an asset (usually a company) in another country (Akhtar et al. 2024). In FDI parlance, the investors are referred to as coming from the "home" country or state, while those from the state being invested in are the "host." Investors do this to obtain access to land resources (including mines, agricultural land, and real estate) or to obtain access both to labor and markets. Countries seek out FDI because it can increase tax revenue, create jobs, encourage technology transfer, and other positive benefits for the economy. Given that "[a]lmost every developed country has had the assistance of foreign finance to supplement its own meager savings during the early stages of its development," those countries who themselves are developing are thus prompted to adjust their economic policies specifically to attract FDI (Lokesha and Leelavathy 2012, p. 459).

The problem is not so much with trade itself, or the fact that trade has become so much more globalized and interconnected, but rather with the way international trade is *governed*—the rules underpinning FDI. These are established by what are known as International Investment Treaties (IITs). These are necessary because a mechanism is needed for establishing the rules and procedures between at least two different foreign governments that do not, in all likelihood, share a common legal or judicial system. Whereas these provisions were once largely established by contract between the investor and the host state, these are now overwhelming provided by treaty (Voss 2010). It is these treaties, nearly 2500 currently in effect worldwide, negotiated either on a bilateral or multilateral basis, which are the problem (UN Trade and Development [UNCTAD], n.d.-a).

Again, the problem is not with the treaty system per se. Rather, it is with the terms of those treaties, and specifically the series of protections they provide to home-state investors and the mechanism they provide for adjudicating disputes about those protections: Investor-State Dispute Settlement (ISDS).

10.3.3 Understanding the Protections Provided to Investors in IITs

IITs have a very specific purpose: to provide a series of key protections to home-state investors who invest abroad (Akhtar and Weiss 2013). West Germany and Switzerland led the first wave of bilateral investment treaties with West Germany concluding a treaty with Pakistan in 1959. The West German government was particularly sensitive to the need to protect its foreign investments, given its wartime experience with the seizure of German assets around the globe and the realization that the growth of its postwar economy would be overwhelmingly export driven (Brower 2023). The fact that powerful economic actors want to protect their investments abroad is nothing new. "[S]ince its inception, the protection of foreign property laid at the heart of public international law" (Voss 2010, p. 1). What is different today is the extent to which "[d]irect investments in developing countries are overwhelmingly governed by Bilateral Investment Treaties (BITs)" (Elkins et al. 2006, p. 812).

And what do these treaties do? They provide a series of key protections to home-state investors who invest abroad. At their most basic level, "BITs are agreements establishing the terms and conditions for private investment by nationals and companies of one country in the jurisdiction of another [covering] four substantive areas: FDI admission, treatment, expropriation, and the settlement of disputes" (Elkins et al. 2006, p. 812). In practice, these treaties provide investors with three broad categories of protection: (1) protection from expropriation, (2) fair and equitable treatment, and (3) protection from discrimination and disparate treatment.

Of course, the question that flows from this is, who are these foreign investors being protected *from*? It is the governments and regulating authorities of the host state. These treaties, "by their unique combination of features, give foreign investors the most powerful protections, by far, of any private actor in international law" (Van Harten 2020, p. 56). They effectively work as a shield against *any and all* laws or regulations of the host governments that implicate any of these protections—which themselves are very broadly interpreted (Arpoulet and Laribi 2024). These treaties are designed to protect the *investment*, and thus the interests of the home state, often at the expense of the peoples and institutions of the host state "establish[ing] these exceptional protections for foreign investors without creating corresponding responsibilities for them to respect a country's laws and people" (Van Harten 2020, p. 56).

From an integral ecology perspective, the problem with these treaties is that the host states, particularly those in the Global South, agree to them under duress. The need for foreign investment capital is such that governments are willing to incur the costs to their national sovereignty that flow from agreeing to BITs in terms of decision-making and enforcement (Elkins et al. 2006, p. 812). While it is clear that, absent such treaties, foreign investors are much less inclined to invest in a particular country, the practical benefits to the host government for making such agreements is open to debate (Falvey and Foster-McGregor 2018). While not mentioning trade treaties specifically, *Laudato Si'* does call out the fact that "some business groups can come forward in the guise of benefactors, wield real power, and consider themselves exempt from certain rules" (Francis 2015, sec. 197).

However, by themselves, the treaty protections are simply words on paper. It is the enforcement mechanism which is built into BITs–the way which disputes are adjudicated as to whether a state is infringing one of these protections–that is particularly problematic. This brings us to the system of Investor-State Dispute Settlement (ISDS).

10.3.4 How Disputes About These Treaty Protections Are Adjudicated—The ISDS Process

Contracts of any stripe, treaties included, very often include clauses related to dispute resolution, i.e., what to do in the (likely) event that the parties disagree on any aspect related to the performance of the agreement. These can specify which particular body of law applies, as well as the steps that an aggrieved party must take before initiating any kind of action. In the absence of this language, such disputes would likely be heard in the national courts of one of the parties, with the rules of contract interpretation, as well as remedies available, specified by that system of law. However it is specifically to avoid the risk of having their disputes heard in the courts of the developing world, which in some cases were former colonies or newly independent countries, that BITs established Investor-State Dispute Settlement (ISDS) as a *private* means of arbitrating disputes in the perceived name of impartiality (Arpoulet and Laribi 2024; Van Harten 2020).

ISDS is a private arbitration system, outside the purview of any national court system, designed to adjudicate foreign investors' claims that host states are in violation of treaty provisions (Van Harten 2020). The phrasing of that statement is not by chance. It is the investors, and *only* the investors, which have the right to initiate an ISDS claim. There is no reciprocal right for a state to initiate a claim against an investor (Arpoulet and Laribi 2024).

An ISDS action generally begins when an aggrieved investor, either an individual or a corporation, initiates a demand for arbitration pursuant to the rules of the International Center for Investment Disputes (ICSID), which is the primary forum for these claims and is affiliated with the World Bank (Murrill 2015). ICSID was established by the ICSID Convention, which came into force in 1966 as a means of creating the framework for adjudicating investment disputes, providing the procedural rules for arbitrations under its auspices (Baltag 2016). Although the parties may agree to hold ISDS proceedings in any location, the vast majority of these proceedings take place in the Global North, either at ICSID's headquarters in Washington or in Paris (International Center for Investment Disputes [ICSID], n.d.). While the parties may agree that an alternative set of rules applies to an ISDS arbitration, such as those of the United Nations Commission on International Trade Law (UNCITRAL), most

BITs negotiated since the end of the 1970s call for the use of ICSID rules (St. John 2018; Taylor 2023).

While there are certain key differences in ICSID and UNCITRAL rules–not to mention those which apply to other arbitral forums–for our purposes it is ISDS on a *structural* level which concerns us. This is because there are problems inherent to the way that ISDS works from a macro perspective that may be nuanced by the application of one set of arbitral rules over another, but that are never entirely mitigated. Among these are questions related to several factors, including: (1) the independence and impartiality of the arbitrators, (2) the transparency of the proceedings, (3) the "early dismissal of frivolous claims," and (4) the ability of third parties and civil society to participate in the proceedings (Murrill 2015, p. 6).

This third category of Murrill's, regarding "the early dismissal of frivolous claims," merits further attention because it directly points to the regulatory chilling effect that the ISDS process works on Global South governments. According to Murrill, an ISDS claim is frivolous if it is "clearly insufficient on its face" and is "presumably interposed for mere purposes of delay or to embarrass" (Murrill 2015, p. 11). He perceives this as a problem given the fact that, even for a manifestly frivolous claim, a host state can spend millions of dollars in defense and attorneys' fees (Murrill 2015, p. 12). While this is true, it points to a much deeper problem with the costs inherent in defending an ISDS claim: the risk of crippling monetary awards made by these private tribunals against host states, which are then, in turn, enforced in the courts in the Global North and can result in the seizure of sovereign assets held abroad (Lemoine 2024).

ISDS tribunals do not generally mandate specific performance when they find in favor of an investor, i.e., they do not require host states to act in a certain manner. The monetary damages which they do award work in largely the same fashion to compel state behavior (Columbia Center of Sustainable Investment 2022). Multiple studies have shown a sharp increase in the average award size over the last thirty years, rising from averages in the tens of millions of US dollars at the turn of the century to the hundreds of millions today (Bonnitcha et al. 2023). These are not trivial amounts, particularly for states which are already under severe fiscal and budgetary constraints, not to mention those in severe debt distress (Calvert 2022). The net result of being so dependent on a system of international trade that has ISDS at its center, and the exorbitant financial risk that attaches to running afoul of foreign investors, is that the arbitration mechanism enshrined in BITs makes host states regulatory averse. In simple terms, ISDS has a chilling effect on the willingness of states to act or legislate in a manner which could invite arbitration claims, *irrespective* of the hopes, needs, or desires of their national populations.

According to Lemoine, "certain sovereigns dominate the world of finance and seem more sovereign than the others" (Lemoine 2024, p. 13). In the case of ISDS, it is *private* foreign investors who have increasingly become "more sovereign" than the actual sovereign and have appropriated for themselves an effective veto over the governments' freedom of action. Where this has particular importance from an integral ecology perspective is in how ISDS has a direct impact on the ability of states in the Global South, particularly in Africa, to govern on climate issues.

10.3.5 How ISDS Affects the Ability of African States to Govern on Climate Issues

According to UNCTAD, FDI to developing countries in 2023 fell 7% to approximately $867 billion, of which $53 billion flowed into Africa (UNCTAD, n.d.-b). Although that number seems relatively modest in comparison to the rest of the developing world, in Africa FDI represents a larger proportion of GDP. Trade remains the gateway to both the global economy and the opportunities for growth and development which are perceived to flow from openness to trade (Arthur et al. 2024). However, as we have seen, the price of entry for states to the global trade system is to subject oneself to the normative framework enshrined in the system of international investment treaties, particularly that of ISDS.

ISDS, as we will recall, is designed to adjudicate disputes as to whether a host state is infringing a foreign investor's treaty protections in three broad categories: (1) protection from expropriation; (2) fair and equitable treatment; and (3) protection from discrimination and disparate treatment (Van Harten 2020). The nuances of how these claims get pled and argued from a legal perspective is not so important for our purposes, but rather the fact that BITs and the ISDS mechanism "allows foreign investors to sue host governments for government measures that undermine their profitability, no matter the objective of the challenged measure" (Sachs et al. 2022, p. 316).

For example, as Sachs, Merrill, and Johnson note, "[i]nvestors have increasingly—and overwhelmingly—used the ISDS mechanism to challenge government measures of general applicability, including those taken in the public interest to protect the life, health, and security of people or the planet, and have been awarded substantial sums by the tribunals, out of public coffers, to compensate investors for economic impacts on their investments, including, in many cases, future lost profits from an impacted investment" (Sachs et al. 2022, p. 317). This puts a tremendous amount of power in individual foreign investors and corporations, neither of which are accountable to Global South electorates, to compel governments to act, or refrain from acting, in ways that they feel are detrimental to their investments. Chinese investors, for example, have recently begun to challenge national security decisions of host states before ISDS panels, arguing that measures designed to restrict Chinese access to technology on military or security grounds implicates treaty protections (Du 2022).

Moreover, from a climate perspective, the character of the investment itself has no bearing on the protections afforded under a BIT. Traditional international investment agreements "do not distinguish between climate friendly and unfriendly investments" (Akinkugbe and Majekolagbe 2023, p. 171). Thus, attempts to mitigate or adapt to climate change are stymied on account of the fact that *any* foreign investment, irrespective of its environmental impact, is entitled to treaty protection. Therefore, a mitigation measure which can be characterized (quite rightly) as seeking to discriminate against a polluting enterprise or investment can, and does, serve as the basis for an ISDS claim (Asenso 2024). Moreover, it is not entirely clear what ISDS panels should do in cases where investments are tainted by corruption, or by what standards those allegations are sustained, meaning that a *prima facie* corrupt investment may be the basis of an ISDS claim (Brewster 2023).

Compounding this problem is the fact that there are no rules to prohibit the buying and selling of international investment claims, in essence allowing third parties to "purchase" the right to initiate and pursue an ISDS claim against a sovereign state (Claussen 2020). This leads to the practice of "treaty shopping," which "generally refers to a practice in which an investor (typically, a multinational corporation) attempts to benefit from more favorable substantive and procedural rules in a particular IIA by acquiring, establishing, or using an existing subsidiary in order to bring an ISDS claim against a host country under that IIA" (Murrill 2015, p. 5). Thus, if a corporation is inclined to put pressure on a particular state, it can structure an investment to purchase a subsidiary which would give it the right to bring a claim under an IIA where they might not otherwise have the ability. Nor is this something that only occurs in the abstract. Colombia is currently in the process of trying to renegotiate its 2014 BIT with the United Kingdom precisely on the grounds that the ISDS process unduly favors multinational corporations over the state (Daniels 2024).

The ability of national courts to intervene in, or review the decisions of, ISDS panels is tremendously limited (Pahis 2023). This is the case even if claims implicate substantive rights or constitutional matters (Linares-Cantillo 2021). The panels themselves have a tremendous amount of discretion and are not bound by precedent, either their own or those of other panel's ruling under a similar set of facts. This leads to investment disputes being adjudicated in an inconsistent, unpredictable, and incoherent manner (Calvert 2022). Recognizing the flaws–specifically the lack of transparency and accountability in the system–the European Union has moved to end the ISDS system and the use of BITs among member states, but has placed no similar prohibition on members entering such agreements with non-EU states (Collet 2021).

10.3.6 How the Regulatory Chilling Effect Works in Practice in Africa

Africa is disproportionately burdened by the effects of climate change: not only the increasing incidences of extreme drought and flooding, but also increases in temperature across the continent (World Meteorological Organization 2024). Seasonal rainfall has fallen sharply, and it has become much more difficult to predict the start of the rainy season, which has a tremendous impact on food security, leading to famine conditions in some areas that are heavily dependent on agriculture (Dunning et al. 2018; World Food Programme 2024). This has a direct impact on those areas that depend on hydroelectric power. Over the last year, we have seen this most clearly in Zambia, where the water level in Lake Kariba, the world's largest dam reservoir, has fallen so low that cities and towns can now go up to three consecutive days without power (Gondwe 2024).

Compounding this crisis is the fact that Africa is being called upon to respond to a crisis largely not of its own making. This is a point not lost on Pope Francis in his 2023 apostolic exhortation *Laudato Deum* (Francis 2023). Africa, according to Jakkie Cilliers, "is a climate change taker, even if it was not a climate change maker," disproportionately vulnerable to its impacts, even if it had little to do with creating the problem (Cilliers 2021, p. 357). In addition to the power generation and food insecurity issues listed above, Africa also has to respond to climate-induced migration, as well as the violence and instability that flows from that (International Organization for Migration 2024). All of these factors mean that African states are being called to respond to, as well as adapt to, climate change at the very moment that many of them are in debt distress and under severe fiscal constraints accelerated by the COVID-19 global pandemic (Arko 2024). However, at the very moment that bold, innovative planning is called for, when the lives of real human beings is at stake, their range of action is chilled by the risk of running afoul of the foreign investors of the past. ISDS and the specter of crippling financial awards paid to the very same industries whose effects they are trying to mitigate hangs menacingly over African governments.

One area in particular where climate mitigation measures and the disinclination of African states to regulate in areas that could provoke ISDS claims interact is in the area of mineral resources. This, again, is problematic from an integral ecology perspective in that it is these same critical minerals, such as cobalt, lithium, nickel, and manganese, which are powering many of the new green technologies which are being touted as sources of clean, renewable energy (Manjonjo and Mokgonyana 2024). While Northern countries tout these minerals as the key to a clean energy transition away from fossil fuels, regulations in the South which seek to manage how fashion in which these minerals are mined, either from an environmental or labor perspective, run the risk of being interpreted as infringing on investors' earlier profit expectations.

Nor is it simply access to this "new generation" of minerals that can provoke investment claims: the fossil fuel industry, and the investments made to further its development, are very much alive. For example, the Energy Charter Treaty, which was negotiated in the 1990s to encourage development of oil and gas resources in the former Soviet Union, specifically protects investments by the fossil fuel industry among member states (Konoplyanik and Wälde 2006). As the Energy Charter Treaty includes a well-used ISDS mechanism, it is not hyperbole that many consider this instrument to be "the most dangerous investment treaty to the energy transition" (Dilworth and Lee 2024).

All of this means that no matter the commitments made by world leaders to reach targets under the Paris Agreement in 2015, or any of the 29 resulting UN Climate Change Conferences (COP) since then, a serious structural barrier in the normative framework of the global financial order exists to hinder their realization. ISDS (particularly in Africa and the Global South) and the risk of crippling monetary awards to aggrieved foreign investors which that system produces work to create a regulatory chilling effect and serve as a disincentive for states to legislate in a manner which may mitigate the effects of climate change. No amount of advocacy, whether at the grassroots or the institutional level, will have any hope of compelling states to act differently unless this structural roadblock is addressed.

10.4 Toward an Integral Ecology—Hope as a Bridge Between Law and Theological Ethics

Hope, ultimately, is the axis on which all of this turns, because without it there is no reason to believe that things can improve. While Christianity has a rich tradition based on the fruit of centuries of reflection on the meaning of hope as well as the *reasons* for hope, but the key point for our purposes here is the firm conviction that human beings are at their essence *good*. This means, in practice, that people can be converted from injurious behavior, that they can be convinced, and that we can do better. The world is not hope-less, but quite the opposite.

This does not mean, however, that the work is not difficult or that change will come about quickly. What integral ecology brings to the table is the idea that there is a spiritual dimension to the ecological crisis, which is no less important than any other factor. Understanding that spiritual dimension recognizes that the difficulty of this work, including the structural and normative obstacles that are firmly ensconced throughout the global economy, can result in a sense of futility and irrelevance to those engaged in it. This is a problem that is just as much spiritual as it is practical.

ISDS is only one example of such a structural barrier, the complexity and pervasiveness of which suggests that there is little reason to hope for change. Part of the reason for this is due to the complexity of the problem at the structural level, given that so much of it takes place in a nontransparent manner in language comprehensible only to lawyers, financiers, and diplomats. Yet, those who profit off of ISDS count on that very complexity and opacity, so that the wider public understand neither the mechanism nor its impacts or results. To counter this phenomenon, there is much that those of us who *do* understand the law (particularly those of us who practice in the Global North) can do to change the system from within by exposing the injustice of the ISDS system. Seeing these links of interconnectedness and the ways that our legal training allows us to be the catalysts for structural change by fostering *dialog* is one of the key insights of Pope Francis's integral ecology. As he writes, "[t]oday in view of the common good, there is an urgent need for politics and economics to enter into a frank dialog in service of life, especially human life" (Francis 2015, sec. 189). "All it takes," the pope reminds us, "is one good person to restore hope" (Francis 2015, sec. 71).

Moreover, because the structural critique of ISDS is ultimately a moral and ethical argument, there is a tremendous space for law to be in conversation with theology, as integral ecology posits that these are not two mutually exclusive intellectual domains, but ones with rich interconnections. This is especially the case in that changing these structures calls for *ecological conversion*, a subject that Francis dedicates the entire third chapter of *Laudato Si'* to and which is rich in theological meaning.

For those of us in the North, individual ecological conversion is not enough. It is critical for us to change our individual consumption patterns and reduce our carbon footprints. Nevertheless, without a *structural* conversion of our social relations and the ways power is exercised on the international level, the work will remain woefully incomplete (Renouard and de Bénazé 2023). There are efforts at reform taking place on the international level, but more needs to be done not only to publicize that work but to pressure policymakers in the North to lend visibility and support. According to French theologian Emmanuel Durand, "[integral ecology] is unique in recognizing the intimate connection between ecological conversion, social justice, and a spirituality of creation" (Durand 2024, p. 42). It is this intimate connection that invites law and theology to a much deeper, critically needed dialog aimed toward bold, interdisciplinary, structural thinking on climate. Everything is interconnected, and because of that we have reason to hope.

References

Akhtar SI, Weiss MA (2013) U.S. International investment agreements: issues for congress (CRS Report No. R43052). Congressional Research Service. Retrieved July 29, 2025 from https://crsreports.congress.gov/product/pdf/R/R43052.

Akhtar SI, Cimino-Isaacs CD, Schwarzenberg AB (2024) Foreign direct investment: background and issues (CRS Report No. IF10636). Congressional Research Service. Retrieved July 29, 2025 from https://crsreports.congress.gov/product/pdf/IF/IF10636.

Akinkugbe OD, Majekolagbe A (2023) International investment law and climate justice: the search for a just green investment order. Fordham Int Law J 46(2):169–212

Arko T (2024) Climate finance, debt and economic dependency in Africa. United Nations University, Institute for Natural Resources in Africa. Retrieved July 29, 2025 from https://unu.edu/inra/article/climate-finance-debt-and-economic-dependency-africa

Arpoulet V, Laribi M (2024) Une justice au service des multinationales [Justice at the service of multinationals]. Le Monde diplomatique, 840:19. Retrieved July 29, 2025 from https://www.monde-diplomatique.fr/2024/03/ARPOULET/66651

Arthur B, Saha M, Sarpong FA, Dutta KD (2024) Unlocking Africa's potential: the transformative power of foreign direct investment for sustainable development. Heliyon 10(5):e26507. Retrieved July 29, 2025 from https://doi.org/10.1016/j.heliyon.2024.e26507

Asenso J (2024) Sierra club: ISDS hinders global climate efforts, should be cut from trade deals. Inside US Trade 42(21)

Baltag C (2016) Chapter 1: the ICSID convention: a successful story—the origins and history of the ICSID. In: Baltag D (ed) ICSID convention after 50 years unsettled issues. Kluwer Law International BV, pp 1–24

Bernardi P (2017) Pope Francis and Ignatian Discernment. In: Buddle ML (ed) New world pope: pope francis and the future of the Church. Cascade Books, pp 53–69

Bonnitcha J, Langford M, Alvarez-Zarate JM, Behn D (2023) Damages and ISDS reform: between procedure and substance. J Int Dispute Settlement, 14(2):213–241. Retrieved July 29, 2025 from https://doi.org/10.1093/jnlids/idab034

Brewster R (2023) Arbitrating corruption. Washington University Law Review 101(3):923–960

Brower CH III (2023) Against imperial arbitrators: tthe brilliance of Canada's new model investment treaty. Florida Int Univ Law Rev 17(1):1–117, Retrieved July 29, 2025 from https://doi.org/10.25148/lawrev.17.1.5

Calvert J (2022) The politics of investment treaties in Latin America (1st ed.). Oxford University Press. Retrieved July 29, 2025 from https://doi.org/10.1093/oso/9780198870890.001.0001
Catta G (2019) Catholic social teaching as theology. Paulist Press
de Charentenay P (2008) Vers la justice de l'Évangile: Introduction à la pensée sociale de l'Église catholique. [Towards Gospel Justice: Introduction to the Social Thought of the Catholic Church]. Desclée de Brouwer
Cilliers J (2021) The future of Africa: challenges and opportunities. Springer Nature. Retrieved July 29, 2025 from https://doi.org/10.1007/978-3-030-46590-2
Claussen K (2020) The international claims trade. Cardozo Law Rev 41(5):1743–1803
Collet P (2021) The current European Union investor state dispute settlement reform: a desirable outcome for investment arbitration? New York University J Int Law and Policy 53(2):689–699
Congregation for the doctrine of the faith and dicastery for promoting integral human development (2018) Oeconomicae et pecuniariae quaestiones [Considerations for an Ethical Discernment Regarding Some Aspects of the Present Economic-Financial System]. Retrieved July 29, 2025 from https://www.vatican.va/roman_curia/congregations/cfaith/documents/rc_con_cfaith_doc_20180106_oeconomicae-et-pecuniariae_en.html
Cook AE (1994) God-talk in a secular world. Yale J Law and the Human 6(2):435–462
Czerny M, Barone C (2022) Siblings all, sign of the times : the social teaching of Pope Francis. In: Paparella J (Tran.). Orbis Books
Daniels J (2024) Colombia seeks renegotiation of investment treaty with the UK. Financial times. Retrieved July 29, 2025 from https://www.ft.com/content/6aba7afd-2e7b-4751-ae7a-19bf4f8ea3f2
Deck AF (2022) Pope Francis and the challenge of ecclesial introversion: where is he coming from? Where is he going? Perspectiva Teológica 54(3):703–717. Retrieved July 29, 2025 from https://doi.org/10.20911/21768757v54n3p703/2022
Dilworth J, Lee E (2024) The energy charter treaty remains the most dangerous investment treaty to the energy transition. E3G. Retrieved July 29, 2025 from https://www.e3g.org/news/the-energy-charter-treaty-remains-the-most-dangerous-to-the-energy-transition/
Du M (2022) Huawei strikes back: challenging national security decisions before investment arbitral tribunal. Emory Int Law Rev 37(1):1–52
Dunning CM, Black E, Allan RP (2018) Later wet seasons with more intense rainfall over Africa under future climate change. J Climate, 31(23):9719–9738. Retrieved July 29, 2025 from https://doi.org/10.1175/JCLI-D-18-0102.1
Durand E (2024) Théologie de l'espérance [Theology of Hope]. Éditions du Cerf, Paris
Elkins Z, Guzman AT, Simmons BA (2006) Competing for capital: the diffusion of bilateral investment treaties, 1960–2000. Int Org 60(4):811–846. Retrieved July 29, 2025 from https://doi.org/10.1017/S0020818306060279
Euvé F (2021) Théologie de l'écologie : Une Création à partager [Theology of Ecology: A Creation to Share]. Éditions Salvator, Paris
Falvey R, Foster-McGregor N (2018) North-South foreign direct investment and bilateral investment treaties. World Economy, 41(1):2–28. Retrieved July 29, 2025 from https://doi.org/10.1111/twec.12539.
Flournoy AC (2020) Integrative environmental law: a prescription for law in the time of climate change. Duke Environ Law and Policy Forum 30(2):225–292
Francis & Ivereigh A (2020) Let us dream : the path to a better future. Simon & Schuster
Francis (2015) Laudato Si' [On Care for Our Common Home][Encyclical letter]. Retrieved July 29, 2025 from: https://www.vatican.va/content/francesco/en/encyclicals/documents/papa-francesco_20150524_enciclica-laudato-si.html
Francis (2023) Laudato Deum [On the Climate Crisis][Apostolic Exhortation]. Retrieved July 29, 2025 from: https://www.vatican.va/content/francesco/en/apost_exhortations/documents/20231004-laudate-deum.html
Gondwe K (2024) How a mega dam has caused a mega power crisis for Zambia. BBC. Retrieved July 29, 2025 from https://www.bbc.com/news/articles/cx2krr137x9o.

Van Harten G (2020) The trouble with foreign investor protection (1st ed) Oxford University Press

International Centre for Settlement of Investment Disputes (nd) A guide to ICSID's global facilities. Retrieved December 18, 2024 from https://icsid.worldbank.org/news-and-events/speeches-articles/guide-icsids-global-facilities

International Organization for Migration (2024, 26 March) Conflict drives displacement amidst rising climate shocks: New Africa Migration Report. Retrieved July 29, 2025 from https://www.iom.int/news/conflict-drives-displacement-amidst-rising-climate-shocks-new-africa-migration-report

Jamieson D (2015) Why Laudato si' matters. Environ: Sci Pol Sustainab Dev 57(6):19–20. Retrieved July 29, 2025 from https://doi.org/10.1080/00139157.2015.1089140

Columbia Center on Sustainable Development (2022, January) Primer on international investment treaties and investor-state dispute settlement. Retrieved July 29, 2025 from https://ccsi.columbia.edu/content/primer-international-investment-treaties-and-investor-state-dispute-settlement

John Paul II (1987) Sollicitudo rei socialis [The Social Concern][Encyclical letter]. Retrieved July 29, 2025 from: https://www.vatican.va/content/john-paul-ii/en/encyclicals/documents/hf_jp-ii_enc_30121987_sollicitudo-rei-socialis.html

Konoplyanik A, Wälde T (2006) Energy charter treaty and its role in international energy. J Energy Nat Resources Law. 24(4):523–558

Lemoine B (2024) Chasseurs d'États : Les fonds vautours et la loi de New York à la poursuite de la souveraineté [State hunters: Vulture funds and New York law in pursuit of sovereignty]. Éditions La Découverte, Paris

Linares-Cantillo A (2021) The intervention of constitutional courts in international investment law: the case of Colombia. Cornell Int Law J 54(1):47–96

Lobez F (2022) Credo et crédit: La pensée sociale catholique et la finance [Creed and credit : Catholic social thought and finance]. Les Belles Lettres, Paris

Lokesha BK, Leelavathy DS (2012) Determinants of foreign direct investment: a macro perspective. Ind J Ind Relat 47(3):459–469

Manjonjo K, Mokgonyana K (2024) Foreign countries are lining up to exploit Africa's critical minerals. Africa at LSE. Retrieved July 29, 2025 from https://blogs.lse.ac.uk/africaatlse/2024/10/28/foreign-countries-are-lining-up-to-exploit-africas-critical-minerals/

Murrill BJ (2015) Issues in international trade: a legal overview of investment-state dispute settlement (CRS Report No. R43988). Congressional Research Service. Retrieved July 29, 2025 from https://crsreports.congress.gov/product/pdf/R/R43988

Obstfeld M, Taylor AM (2017) International monetary relations: taking finance seriously. J Econom Perspect 31(3):3–28. Retrieved July 29, 2025 from https://doi.org/10.1257/jep.31.3.3

Pahis S (2023) Rethinking international investment law: form, function and reform. Virginia J Int Law 63(3):447–495

Ramirez SA (2017) Social justice and capitalism: an assessment of the teachings of pope francis from a law and macroeconomics perspective. Seattle Univ Law Rev 40(4):1229–1259

Renouard C, de Bénazé X (2023) Rouvrir l'horizon: Manifeste d'espérance engage face aux effondrements [Reopening the horizon: A manifesto of hope in the face of collapse]. Éditions Emmanuel, Paris

Rowlands A (2021) Towards a politics of communion: catholic social teaching in dark times. T&T Clark, London

Sachs L, Merrill E, Johnson L (2022) Environmental injustice: how treaties undermine the right to a healthy environment. In: Sachs JD, Sánchez Sorondo M, Flanagan O, Vendley W, Annett A, Thorson J (eds) Ethics in acton for sustainable development. Columbia University Press, pp 316–326

Setser B (2024) The dangerous myth of deglobalization: misperceptions of the global economy are driving bad policies. Foreign Affairs. Retrieved July 29, 2025 from https://www.foreignaffairs.com/china/globalization-dangerous-myth-economy-brad-setser

St. John T (2018) The rise of investor-state arbitration : politics, law, and unintended consequences (1st ed.). Oxford University Press

Taylor T (2023) Much dispute about nothing? A critical examination of the backlash against investment treaty arbitration in international intellectual property disputes. Cybaris 14(1):33–70

Tooze A (2019) Crashed: how a decade of financial crises changed the world. Penguin Random House UK, London

Tooze A (2021) Shutdown: how covid shook the world's economy. Viking

UN Trade and Development (nd-a) Investment policy hub: investment agreements navigator. Retrieved December 11, 2024 from https://investmentpolicy.unctad.org/international-investment-agreements

UN Trade and Development (n.d.-b) World investment report 2024. Retrieved December 19, 2024 from. https://unctad.org/publication/world-investment-report-2024

Voss JO (2010) The impact of investment treaties on contracts between host states and foreign investors. Brill

World Food Programme (2024, 22 May) Devastating drought and floods in Southern Africa: WFP Chief calls for global action as millions face food insecurity. Retrieved July 29, 2025 from https://www.wfp.org/news/devastating-drought-and-floods-southern-africa-wfp-chief-calls-global-action-millions-face

World Meteorological Organization (2024, 2 September) Africa faces disproportionate burden from climate change and adaptation costs. Retrieved July 29, 2025 from https://wmo.int/news/media-centre/africa-faces-disproportionate-burden-from-climate-change-and-adaptation-costs

World Trade Organization (2024) World trade report 2024: trade and inclusiveness. How to make trade work for all. Geneva: WTO Publications. Retrieved July 29, 2025 from https://www.wto.org/english/res_e/booksp_e/wtr24_e/wtr24_e.pdf

Fernando C. Saldivar, S.J. Jesuit priest and attorney whose work focuses on global justice, ethics, and sustainable development. His work has particularly addressed climate justice, debt, taxation, and social development in Africa through his engagement with the Jesuit Justice and Ecology Network Africa (JENA). He serves as Special Assistant to the Dean and Affiliate Faculty in Human Rights at LMU Loyola Law School in Los Angeles, California.

Chapter 11
Making Hope Possible Through "Service": The Case of the Roman Dicastery for Human Development

María Dolores Sánchez Galera

Abstract The ecological crisis stems from various fundamental human conditions. These can be analyzed through the lenses of the "integral ecology" paradigm which has been developed by the Catholic Church and its roman curia. It combines spiritual and moral concerns with a plentiful praxis of hope in the modern world. The Dicastery for integral human development of the curia promotes this paradigm, especially inspired by the magisterium of Pope Francis' Pontificate. This allows us to better understand the necessity of key actions and multiple alliances for a praxis of hope in the present.

Keywords Integral human development · Promotion · Institutional praxis of hope · Catholic Church · Roman Curia

11.1 Human Foundations of the Ecological Crisis

The Catholic Church has always rooted its tradition of socio-economic thought in the complexity of the human being in its relationships with oneself, with others, with nature, and with God who, before anything else, is experienced as transformative in its essence.

Pope John XXIII (1881–1963) was the first to introduce the idea of "integral development of the person" in the encyclical letter *Mater et Magistra* (John XIII, P. 1961). He taught about the need for "Christian education" to be "integral" and encompass every kind of duty. This meant that a Christian in all the areas of his life, whether at work, in family, or as a parent, or in the fields of economics or politics,

M. D. S. Galera
Dicastery for Integral Human Development, Holy See, Rome, Italy

M. D. S. Galera (✉)
Environmental and Sustainability Studies, Pascual Madoz Institute, Universidad Carlos III, Madrid, Spain
e-mail: mariadsa@inst.uc3m.es

T. Karcher SJ et al. (eds.), *Making Hope Possible*, Environmental Humanities: Transformation, Governance, Ethics, Law 7,
https://doi.org/10.1007/978-3-032-17719-3_11

was called to be a responsible citizen and to participate in social activities. The term "integral" was developed further by Jaques Maritain's writings on integral humanism (Maritain 1968) and his work has been highly influential in humanistic and Catholic thought (Maritain 1947). Generally, in Catholic social teaching, integral development has always been connected to the realization of one's potential. Pope John Paul II talks about integral development of the human person through work, for example (1991, 4). It was the Vatican Council II that inspired a more committed study of the relationship between man and his environment,[2] highlighting the importance of the "relational" paradigm as a crucial heritage of Catholic social teaching (from now onwards CST).

Pope Francis (1936–2025) indeed missed no opportunity to remind us that our lives flow through relationships. Everyone's existence is linked to that of others. He always emphasized that life is not time passing, but a time of encounter. The *Laudato Si'* encyclical (Francis 2015) (from now onwards LS) and the *Laudate Deum Exhortation* (Francis 2023) help us to draw our attention to the human roots of the ecological crisis and the urgent need to react. Pope Francis reveals to us the betrayal of the social pact of human relationships, the betrayal of the relationship of humans with nature, and the betrayal of humans' relationship with God. With *Fratelli Tutti* (Francis 2020), this Pope reminded us of the failure of our International institutional setting to face common challenges and the consequences of bad politics. It was the first time that an encyclical letter, which is an authoritative text written by the Pope, addressed matters of doctrine, morals, and discipline within the Church while also touching on broader social issues and even devoting one entire chapter to politics (bad politics in this case). This encyclical also talks of hope and its collective nature.

The *Laudato Si'* encyclical takes its name from the invocation of St. Francis of Assisi as the example *par excellence* of care for the vulnerable and of an integral ecology lived out joyfully and authentically. It reminds us of the centrality of the interconnectedness of human strengths and fragilities, our commonalities supporting subsidiarity and common good. It is and intends to be inspirational, inviting us to contrast technological challenges with moral concerns that can further be translated into spiritual questions.

The integral ecology approach fostered by Pope Francis's *Laudato Si'* encyclical (2015) has put forward evidence of how influential and inspirational the CST can be today in a world thirsty for hope. In response to the complex and interconnected crisis that we all collectively face. More than that, it has fostered a second *momentum* for the action-based and transformative approach of "integral human development" by putting environmental issues in their economic and social context. By promoting an integral view of the world and by reflecting on authentic development explicitly, LS has contributed to a deepening of the concept of integral human development, first introduced by Pope Paul VI, in the encyclical *Populorum Progressio* (1967, 14, 42) teaching that "authentic development must foster the development of each man and of the whole man." LS includes an ecological and long-term perspective that considers future generations and integrates the ineludible notion of intergenerational solidarity that can no longer be disconnected from the context of ecological justice

(Francis 2015, 159). It is an action-based and transformative approach of "integral human development"[4] in so far as it places environmental issues in their economic and social context. By promoting an integral view of the world and by reflecting on authentic development explicitly, *Laudato Si'* has contributed to a deepening of the concept of Integral Human Development with an ecological and long-term perspective that considers future generations and integrates the ineludible notion of intergenerational solidarity that can no longer be disconnected from the context of ecological justice (Francis 2015, 159).

Understanding how the church operates within this context is critical in a world in search of alliances, institutional reforms, and new paradigms that can disrupt the inertia of the economic, social, and political architecture and thereby lead to real transformation. Within this context, it is important to highlight the impact and the influence of Pope Francis' *LS* encyclical (2015) on sustainability, science, and knowledge production (Landrigan et al. 2024; Molina and Perez-Garrido 2022). We will not deal with this here, but the systemic and holistic approach of the moral and spiritual questions dealt with in the text remind us of the complexity of the relational world and help us to better understand the roots of the current ecological crisis. Pope Francis' LS encyclical (2015) analyzes well our complex and deteriorated relationship with others and the ways that this lack of space for "human encounter" triggers catastrophic consequences for our common home. The teachings of this encyclical have shed light on current scientific research in sustainability science and transformational studies.

In 2020 Pope Francis addressed to the Amazonia an inspiring and motivating message, an exhortation, *Querida Amazonia*. There, Pope Francis (2020, n. 58) stated, "an integral ecology cannot be content simply with fine-tuning technical questions or political, juridical, and social decisions. The best ecology always has an educational dimension that can encourage the development of new habits in individual groups." Throughout all his pontificate, Pope Francis speaks of the crises and effects of modern anthropocentrism, suggesting that it "(…) has paradoxically ended up prizing technical thought over reality" (Francis, P. 2015, 115), "(…) leading to a constant schizophrenia, wherein a technocracy which sees no intrinsic value in lesser beings coexists with other extremes, which sees no special value in human beings. But one cannot prescind from humanity. There cannot be renewal of our relationship with nature without a renewal of humanity itself. There can be no ecology without an adequate anthropology" (Francis, P. 2015, 118). These paragraphs are related to specific solutions. The pope shared with the world in his LS encyclical a strong message, stating that "we cannot presume to heal our relationship with nature and the environment without healing all fundamental human relationships" (Francis 2015, 119). This "relational" component is seen as essential to establish fruitful cultural and interfaith bridges to build a different future for humanity, and the institutional reach of the church is crucial for operations in service of this call.

This is in sharp contrast with the technocratic paradigm that seems to master every possible scenario today, and which conceals the necessary complexity of everyone's life. The temptation of responding to complex social challenges through technology and more knowledge is more pervasive than ever. Integral human development reminds us of other fundamental questions that need more courageous and perhaps more difficult responses invoking a change of heart, conversion, and long-term results.[5]

Integral human development, originally presented as a valid paradigm within the context of human flourishing, goes back to Pope Paul VI in 1967. In his encyclical *Populorum Progressio* (from now onwards PP), he talked about "authentic human development," development that "cannot be restricted to economic growth alone. To be authentic must be well rounded; it must foster the development of each person and of the whole person." In line with these indications that have accordingly been discussed in further Church Documents,[6] we cannot deny that this operational concept and its human dignity-based focus is still a valid, inspirational, as well as aspirational approach committed to connecting the gap between our technological possibilities and our moral and spiritual ones.[7] There are five important features that render this concept more accessible to be "measured" at the empirical level, and these have been well analyzed by Sedmak (2023) to be applicable at the time of projecting and accompanying challenging ventures and processes in specific fragile communities. The concept incorporates many of the features of both sustainable and rights-based development, but it has a very different background based on a Catholic view of human anthropology and values intrinsic to the human person, always perceived in relationship with others and bringing in the richness of unique social and cultural contexts. The concept thus advances a more rounded view of what development means.

It is therefore important to place integral human development in its full context within the Catholic Social Thought (CST) tradition. This tradition includes the magisterial documents containing CST principles but also the works of those thinkers who can complement Papal documents with other perspectives based on the same moral foundations. This is necessary partially because there is no comprehensive teaching on integral human development per se and because it includes the experience of the institutions of the Roman Curia, the praxis which is inspired by the magisterium, together with the action of development agencies outside the catholic world, and the voices of those from differing cultural contexts in the contemporary world, especially those from outside Europe and North America. They can offer new ideas drawn from social sciences, contextual theologies, and experiential learning. CST is not built on a certain developmental theory. Rather, it represents a reflection on the social situation from the perspective of moral theology and philosophy and is offered as a guide in our search for new roads to development for transformation at the Universal level that place the dignity of man at its center.

11.2 The "Service" of a Dicastery of the Roman Curia

The mission of the Dicastery for the Promotion of Integral Human Development (henceforth DPIHD) is specified by the new Apostolic Constitution *Praedicate Evangelium* (2021), from now onwards PE. The Apostolic Constitution sets the basic institutional principles of the Roman curia. The Roman Curia serves as the central administrative body of the Catholic Church, assisting the Pope in his role as the head of the Church and in his universal mission. Such missionary action today is essential within the context of hope, striving for social and ecological justice and removing obstacles to human development. In the PEs' preamble 8, the importance of service for the Roman Curia is specified. The main purpose of the Roman Curia's institutions and bodies is to support the Pope, the Bishops (broadly understood), and the faithful in promoting integral human development, ensuring that no one is excluded. The specific competences associated to the DPIHD's mission of the Roman Curia are enshrined in arts. 163–174 PE: "[I]n cooperation with Episcopal Conferences, their regional and continental groupings and the hierarchical structures of Eastern Churches, the Dicastery fosters the implementation of the Church's magisterium in the areas of the protection and integral development of the environment, working with members of other Christian confessions and of other religions, with civil authorities and organizations, and with international organizations" (art. 164 PE) when needed. It is an action-based approach that presupposes knowledge, but it is guided by the perspective of people's dignity, which is also specified by art. 165 PE, and their freedom to live good lives, "life in abundance." Art. 163 PE specifies the mandate entrusted to the DPIHD with the task of "(…) promoting the human person and the God-given dignity of all, together with human rights, health, justice, and peace." It is principally concerned with matters relating to the economy and work, the care of creation and the earth as our "common home," and migration and humanitarian emergencies.

The strength of the DPIHD action paradigm lies in its attention to and focus on the margins (Sedmak 2023, 72). Drowning in deep acknowledgment of today's rising inequalities, it takes care of the "cry" of the most vulnerable populations. It has a bottom-up approach implemented through an institutional organization focused on service and the accompaniment of local churches. It identifies worldwide fragilities and empowers communities rather than isolating individuals to contrast them. Mission and service are the key aspects of this Dicastery, concretized by deep listening and the accompaniment of Bishops, local churches, congregations, and other actors all over the world in their hard and solitary mission to render good service to their communities.

How does this service become concrete? The approach is at first based on listening. Deep listening becomes key to building a relationship with communities and fruitful dialog for long-lasting action and robust connections. Let us say that listening creates the necessary conditions for the experience of dignity. This is also emphasized in the way one pays specific attention to the margins, to what Pope Francis calls "existential peripheries," the most vulnerable people, the ones who suffer the most. Listening to them face-to-face. This precludes a predetermined agenda. Specific empirical research conducted on governance arrangements of common-pool resources have shown how face-to-face communication increases the ability of a community to properly manage such common resources (Ostrom 1998, 9–10). Within the Dicastery experience, this is a two-way bridge with local churches and its various ministries promoting integral human development.

Sedmak (2023), in his excellent work "Enacting Integral Human Development," argues with some evidence how the art and skill of listening emerges as a major virtue for dignity-based project management. He explains well how there is an understanding of persons, but a relational understanding of development of the human person, taking care of three important elements: encounter, presence, and accompaniment (Sedmak 2023, 97–103).

The second stage of the Dicastery's action is centered on reflection on people's sufferings, identifying priorities, and choosing the right action in the face of complex challenges. This offers the Dicastery the possibility to understand how complex situations are often accompanied by hopeful processes of action and accompaniment that favor collaboration between different actors. Thus responding to the needs of local Churches facilitates collaboration with communities, experts, research institutions, organizations, etc., and helps to deepen and spread CST. Everyone who is affected by a specific challenging circumstance is at the center of the Church's support with different ways of accompaniment. But not a single objective is set in advance. We are nourished by relations and people's needs as they face real obstacles to a dignified life—not just refugees, migrants, mothers, workers, or youth. The work of the Dicastery consists of identifying their priority requests, understanding the obstacles that hamper their proper flourishing, and accompanying them in the development and use of appropriate pastoral tools.

The third stage consists of restitution through communication and dissemination, as both an inspirational action and an opportunity of living the Gospel in concrete instances. The goal is to have an impact beyond the Church institutional setting and context. The emphasis on the centrality of the dignity of all men and women and of those who are especially vulnerable, evidenced by the best practices reported by our communication networks, provides a crucial guide for action with regard to the most disadvantaged as a key figure of Integral Human Development.

Going through a process of transformation, the DPIHD has recently strived to make this recent reform of the Roman Curia by Pope Francis visible after the enactment of the above-mentioned PE. The reform was motivated by Pope Francis' desire "to give the Church a missionary outlook and to make its structures more mission oriented," "[which] is a heartfelt invitation to recover the original spirit of the Roman Curia without renouncing the skills and experience acquired over its centuries of institutional development" (Ekpo 2024). Service and communion are at the center of this reform. It has been argued that Pope Francis gave the Apostolic Constitution its title to emphasize his missionary vision for the Roman Curia, a vision inspired by pastoral conversion and leavened by the action of the Holy Spirit (Ekpo 2024).

The DPIHD was first established in 2017 through the Apostolic Letter *Humanam Progressionem*, issued by Pope Francis, in which he addresses some administrative and pastoral issues. In this letter, integral human development was characterized as a kind of development that "takes place by attending to the inestimable goods of justice, peace, and the care of creation" through the visibility of the most vulnerable. Such attention is imbued with high moral grounds and challenges our attitudes and forms of life in relation with others, with God, and with our mother Earth. This has been noticeably advanced during the current Pontificate.

Today, it brings added value in relation to the ecological crisis and humanity's challenges ahead, because of its universal reach, but more importantly: its bottom-up approach; human dignity-based inclusion of each person and the whole person; special consideration of the least; the wealth of communities' and cultural contexts; and the need to overcome indifference and institutional limitations while finding long-lasting solutions, building networks, and engaging with global and local resources to leverage the wealth of communities.

11.3 Relevance for Transdisciplinary Science

Today's urgent worldwide challenges regarding the human crisis and despair, climate crisis, biodiversity losses, ocean pollution, and water scarcity have required new methodological approaches, new ways of knowledge, and decision-making. This has fostered sustainability science as a transformational scientific field. There is emerging agreement that sustainability challenges require new ways of knowledge production and decision-making. One key aspect of sustainability science, therefore, is the involvement of actors from outside academia in the research process to integrate the best available knowledge, reconcile values and preferences, and create ownership for problems and solution options. Transdisciplinary, community-based, interactive, or participatory research approaches are often suggested as appropriate means to meet both the requirements posed by real-world problems as well as the goals of sustainability science as a transformational scientific field (Lang et al. 2012). However, it is the failure of traditional disciplinary approaches, together with the collapse of the credibility of both development approaches and economic well-being built on the capitalist paradigm, that have led to the leveraging of the action-based approach of the Church, based in knowledge and practice, to face complex or wicked problems and eventually foster long-lasting transformation.

There is enough scientific evidence to affirm that the Church has always rooted its tradition of thought in such complexity, thinking and acting beyond and across disciplines for the sake of the common good and building bridges between human needs and institutions, empowering people (subsidiarity principle) and privileging bottom-up approaches.

Pope Francis reminds us in his encyclical *Fratelli Tutti* that "some economic rules have proved effective for growth, but not for integral human development." Wealth has increased, but together with inequality, with the result that "new forms of poverty are emerging" (Francis 2019, 21). It is within this context that the mission of the Church and its projects become crucial for an ongoing institutionalization of integral human development that is consistently operationalized by the Church, engaging knowledge and crucial actors in many ways on a transdisciplinary basis.

The concrete operational experience of the Dicastery for the Promotion of Integral Human Development specified by the PE art. 164 is a good example of how a collaborative process emerges through integrating knowledge and expertise from individuals or disciplines at the service of the common good. In the case of ecological concerns, the cooperation takes place with Episcopal conferences, their regional and continental groupings, and the hierarchical structures of the Eastern Churches. Within such a collaborative framework, the DPIHD fosters the implementation of the Church's magisterium to reflect and help remove specific obstacles that hamper integral development in relation to ecology—working with members of other Christian confessions, other religions, civil authorities and organizations, and international organizations when needed.

This action-based approach presupposes knowledge that goes beyond disciplines. It is guided by the perspective of people and their freedom to live good lives, "life in abundance." It goes to the margins, expressing respect for local agency, represented by the idea of subsidiarity rooted in Catholic Social Tradition. Subsidiarity is presented as a principle that strengthens the agency of individuals and communities (Sedmak 2023, 75).

This bottom-up approach is implemented through an institutional apparatus focused on listening, reflecting, and responding. It is done with the involvement of the Church, civil society, indigenous communities, congregations, governments, and the scientific world, as well as with the social doctrine of the Church, which is always inclusive. Specific results that come out of these collaborative processes refer to Pastoral orientations on issues that concern any sort of human suffering, training activities, educational programs, and specific projects, run together with dioceses or other Church partners. The example of *Casa di Zaccheo* in the Italian City of Mesagne, province of Brindisi, gives some empirical evidence of how the DPIHD and local Churches work together in a synodal way. The Archdiocese of Brindisi, Ostuni, Italy, together with the DPIHD, promoting the project "Your Dignity: from your story and your work." The project seeks to give dignity to people through a tailoring, laundry, and ironing workshop where everyone can carry out activities related to their profession, enhancing their skills and helping them regain personal dignity. Projects are run to become sustainable over time[11]. Associated institutions with a strong impact at local level are Caritas Internationalis and the International Catholic Migration Commission. Many different projects have been developed at the diocesan level (local church jurisdictional context) to deal with the migration problem, including pastoral needs to help the most vulnerable populations and with the fragilities of God's peoples. It is out of the scope of the current contribution to list all the ongoing projects that the DPIHD helps with its universal reach.

11.4 Making Hope Possible?

Transformative experiences for the common good are flourishing all around the world, in the form of action or in the form of knowledge. Within the Catholic faith context, this is an evident result of its renewed and inspiring message. It is a sign that the scientific world started to be influenced consistently through the most recent encyclicals and the socio-economic relevance of their anthropological paradigms—integral human development and integral ecology. The added value of these conceptual approaches lies in their transformative power shedding light into many of the complex social, political, economic, ecological, and moral questions of human life, so that they become inspirational and uplifting beyond the spiritual level.

Creating spaces and opportunities for listening and reflecting involves collaboration and fostering initiatives that bring the social doctrine of the Church into life, transforming specific social contexts. These are initiatives that we generally take for granted within the Catholic operational context. We need to start considering the genuine value and the power of subsidiarity for the common good, accompanying fragile communities and lives. Some initiatives are better known than others (Alkire and Foster 2010). Advocacy and accompaniment are intrinsic to some of the best-known operational bodies of the Catholic Church: *Caritas*, *Catholic Relief Services*, and the work of different Catholic congregations inspired by different charisms that for reasons of space cannot be reported here. Still, we must see that such contributions are not enough in the face of the massive crises through which we are constantly living. Yet the influence of Catholic economic and social thinking on economic history is undeniable (Bucciarelli et al. 2011; Deneulin 2018). Today, the Catholic Social Doctrine is still helping the flourishing of Catholic socio-economic institutions, schools, universities, vocational courses and training institutes, cooperatives, and business organizations, and it is leveraging the strengths of many communities around the world. Some good examples are: the solidarity economy, the economy of communion, refugee services, first aid assistance, diocesan listening centers, generative communities, cooperative movements, agro-forestry initiatives lead by religious networks, urban commons, faith investment initiatives, ecological plans run by congregations, decarbonization projects, cultural and artistic labs integrated into urban settings, the Economy of Francesco, non-extractivist legal approaches, small energy communities, etc. The work of the Dicastery for integral human development establishes synergies and promotes some of these initiatives, either by giving assistance to specific projects or by initiating and accompanying ongoing processes.

Pope Francis has given the theme "Pilgrims of Hope" to the Jubilee of 2025, a holy year, reflecting on the theological virtue of hope. Hope is presented as a gift: not as a mere act of optimism, but as a force that can transform individual and collective reality. This is an idea that has been presented by other contemporary thinkers with the same powerful message. In our technocratic society, "Only hope can give us back that life that is more than mere survival. It is hope that opens a meaningful horizon that reinvigorates and inspires life. Hope presents us with a future" (Han 2024). Han redefines the concept of hope in the challenging context we are facing today, full of crises and catastrophes of ecology (floods, draughts, fires) and problems of human nature (wars). For contemporary authors like Han, true hope is not found in technological advances or in the promises of a future controlled by capital, but in the ability to imagine a different world, a collective vision of a future that escapes the logic of domination and exploitation.

The relevance of Catholic beliefs and their notion of hope in today's world lie in their relational paradigm and their simple, transforming message of love. There is a need for more in-depth research into the relational paradigm that is at the basis of CST. This will transform the current foundations of our socio-economic and political systems so that hope is understood as a necessary collective desire, which endorses solidarity and positive action. Unfortunately, the complexity of the current world fails to gather people around common values, even when dealing with universal concepts such as love, solidarity, and hope.

11.5 Conclusions

Laudato Si', relying on thousands of years of CST, calls on all of us to live out our faith by caring for our common home and by caring for others with perseverance. The interconnectedness and togetherness beyond physical, political, economic, social, or religious bonds is the essence of the integral ecology paradigm introduced by this encyclical, and we should take for granted that it is also at the basis of the notion of hope that flows from the missionary action of the Catholic Church.

Integral ecology is inseparable from integral society, understood as a concrete form of integral human development for which this paradigm, when it is operationalized, leverages people and communities' strengths by fostering processes of human flourishing connected with human dignity lived in well-rounded relationship with others, creation, and moral and spiritual growth.

Despite the level of global awareness, democratic governance, and resources invested for human development, our extractivist economy is sowing seeds of despair in many parts of the world where the voices of many local communities are still unheard (Deaton and Case 2021). Youth are suffering from mental health problems and technological and other types of addiction are creating more social difficulties. All of these people need hope, action, thought, and prayers. The Church, with its institutional bodies and global reach, has always been there, continuously renewing its missionary action when needed while also sharing its simple message of love in action.

References

Alkire, S., & Foster, J. (2010). Designing the inequality-adjusted Human Development Index (HDI), *OPHI Working Papers* 37, Oxford Poverty and Human Development Initiative (OPHI), University of Oxford.

Bucciarelli E, Mattoscio M, Persico T (2011) The Christian ethics of socio-economic development promoted by the Catholic social teaching. J Philosoph Econom 5(1):90–119

Deaton A, Case A (2021) Deaths of despair and the future of capitalism. Princeton University Press

Deneulin S (2018) Integral human development through the lens of Sen's capability approach and the life of a faith community at the Latin American Urban Margins" (Version 1). University of Notre Dame. https://doi.org/10.7274/26125825.v1

Ekpo A (2024) The Roman Curia: history. Georgetown University Press, Theology and Organization

Future of Work 2025 published by the World Economic Forum. Retrieved July 24, 2025, from https://www.weforum.org/stories/2025/01/future-of-jobs-report-2025-jobs-of-the-future-and-the-skills-you-need-to-get-them/

Han, Byung-Chul Han (2024) The spirit of hope. Polity Pr

Lang D et al (2012) Transdisciplinary research in sustainability science: practice, principles, and challenges. Sustain Sci 7:25–43

Laudrigan R et al (2024) Laudato Si' and the emerging contribution of catholic research universities to planetary health. Lancet Planetary Health, 8. https://doi.org/10.106/S2542-5196(24)00012-3

Maritain J (1947) The person and the common good (1st edn) Edition of 1994 published by University of Notre Dame press

Maritain J (1996) Integral humanism, freedom in the modern world, and a letter on independence (Rev. ed., Collected Works XI, R. Royal, Ed.). University of Notre Dame Press. (Original work published 1968)

Molina C, Pérez-Garrido C (2022) Laudato Si' and its influence on sustainable development five years later: a first look at the academic productivity associated to this encyclical. Environ Dev 43:100726. https://doi.org/10.1016/j.envdev.2022.100726

Orson E (1998) A behavioral approach to the rational choice theory of collective action: presidential address, in American political science association, 1997. Am Polit Sci Rev 92(1):1–22

Persico T, Mattoscio M, Bucciarelli E (2011) The christian ethics of socio-economic development promoted by the catholic social teaching. J Philosoph Econom V 1:90–119

Persico T, Mattoscio M, Bucciarelli E (2018) Economics and humanity: theory of development in A. Sen and the concept of human integral development. In: Calvo G, Fiori S, Nishimura B (eds) The individual and the other in economic thought. Routledge, pp 67–79

Pope John Paul II (1981) Laborem exercens [Encyclical letter]. Holy See, Rome. Retrieved July 24, 2025, from https://www.vatican.va/content/john-paulii/es/encyclicals/documents/hf_jp-ii_enc_14091981_laborem-exercens.html

Pope John Paul II (1987). Sollicitudo Rei Socialis, [Encyclical letter]. Holy See, Rome. Retrieved July 24, 2025, from https://www.vatican.va/content/john-paul-ii/es/encyclicals/documents/hf_jp-ii_enc_30121987_sollicitudo-rei-socialis.html

Pope Francis (2015) Laudato Si'. On Care for our Common Home [Encyclical letter]. Holy See, Rome. Retrieved July 24, 2025, from https://www.vatican.va/content/francesco/es/encyclicals/documents/papa-francesco_20150524_enciclica-laudato-si.html

Pope Francis (2020) Fratelli Tutti, [Encyclical letter]. Holy See, Rome. Retrieved July 24, 2025, from https://www.vatican.va/content/francesco/es/encyclicals/documents/papa-francesco_20201003_enciclica-fratelli-tutti.html

Sedmak C (2023) Enacting integral human development. Orbis Books

Zamagni S (2014) Sustainable humanity, sustainable nature: our responsibility. Pontifical academy of science, Extra Series, 4. Vatican City

María Dolores Sánchez Galera Senior Research Advisor for the Dicastery for Integral Human Development in the Holy See and member of the governing body of the Pontifical Academy of Social Sciences. Research Fellow at the Pascual Madoz Institute of Carlos III University of Madrid. Her current work bridges academic research in Law and institutions with the Social Doctrine of the Church, with a particular focus on integral ecology, inequalities and human development.

Chapter 12
Active Hope and Ecological Peace from the Margins: The Role of Higher Education

Mélodie Honen-Delmar, Anna Mayr, and Magdalena Nauderer

Abstract This article examines the transformative role of higher education in promoting ecological justice and social equity through the lens of Active Hope and Barnett's Ecological University framework. Against the backdrop of growing global environmental and social challenges, higher education institutions (HEIs) are presented as increasingly important actors in addressing these challenges. Particular focus is given to JesuitWorldwide Learning's Bachelor of Arts in Sustainable Development (BASD) program, which was developed for marginalized communities, and combines Ignatian pedagogy with experiential and reflective forms of learning. A case study of the Peace Leader program draws on qualitative data from graduates living in refugee camps and underserved regions. Further, it is explored how the program fosters hope, ecological transformation, and peacebuilding at the margins. Lastly, the broader implications for transdisciplinary approaches in higher education to empower vulnerable communities toward sustainable futures and socio-environmental justice are discussed.

Keywords Higher education · Active hope · Blended learning · Ecological justice · Ignatian pedagogy · Sustainable development education · Refugee education · Experiential learning · Peacebuilding · Socio-environmental justice · Marginalized communities

---

M. Honen-Delmar (✉)
Jesuit Worldwide Learning, Geneva, Switzerland
e-mail: melodie.honen-delmar@jwl.global

A. Mayr
Jesuit Worldwide Learning, Uffing, Germany
e-mail: anna.mayr@jwl.global

M. Nauderer
Jesuit Worldwide Learning, Munich, Germany

Jesuit Worldwide Learning, Erbil, Iraq

M. Nauderer
e-mail: magdalena.nauderer@jwl.global

T. Karcher SJ et al. (eds.), *Making Hope Possible*, Environmental Humanities: Transformation, Governance, Ethics, Law 7,
https://doi.org/10.1007/978-3-032-17719-3_12

12.1 Educating for Change: Higher Education and Global Challenges

Higher education has long been recognized as a catalyst for societal change. Over centuries, universities have adapted their roles to advance knowledge and respond to the specific needs and challenges of their times (Acosta Güemes et al. 2023; Barnett 2021; Lorio And Tanabe 2019). In ancient civilizations, such as those of Greece and China, tertiary education focused on philosophy and governance, aiming to train elites and leaders (Acosta Güemes et al. 2023). During the Middle Ages, higher education institutions (HEIs) primarily served to educate clergy and elites in theology, law, and medicine (Miles And Miles 2016; Perkin 2014). By the twentieth century, HEIs underwent a period of diversification and democratization, embracing emerging disciplines like computer science while incorporating new epistemological approaches, including those addressing colonialism, constructivism, orientalism, and gender studies (An 2021; Cairns 2023; Lynch 2022). In the twenty-first century, scholars argued that HEIs must address pressing global challenges, such as sustainable development and social inequalities (Barnett 2021; Cortese 2003; Giroux 2016). Despite their varied purposes across different eras, HEIs have consistently acted as vehicles for change, preparing new generations to fulfill the aspirations of their predecessors. Central to this mission is the concept of hope, which shapes the vision and direction of HEIs.

In this chapter, we propose examining this concept within the sphere of higher education through the framework of Active Hope (Macy And Johnstone 2012), in the context of Jesuit Worldwide Learning (JWL), a global NGO that provides accredited tertiary education to learners in displaced and marginalized communities. By examining JWL's Bachelor of Arts in Sustainable Development (BASD) program and its pedagogical and delivery models, this paper explores the theoretical foundations of higher education as a catalyst for hope, investigating its role in promoting ecological transformation and peace. After introducing the notion of Active Hope (Macy And Johnstone 2012) and Barnett's (2021) model of the ecological university and higher education, this paper brings the perspective of the margins to the center, identifying the need for ecological transformation in marginalized settings. Second, this paper presents the BASD program in light of its Ignatian pedagogy, integrating experiential learning, reflective practices, and meaningful action to understand how higher education can foster a dynamic engagement and empower learners to actively contribute to positive change and sustainable development. Lastly, the impact of the Peace Leader course, a BASD course, is developed to showcase how the BASD program, and higher education in general, can serve as a model for ecological transformation at the margins. Hence, this paper seeks to contribute to the discourse on transdisciplinary approaches to ecological justice, with a focus on the pivotal role of higher education in empowering marginalized learners and fostering hope for socio-environmental justice.

12.2 Theoretical Foundation: Environmental Justice Through the Lens of Active Hope and Higher Education

There is a pressing need for transformation at the margins when considering socio-environmental justice due to the disproportionate impact of environmental degradation and social inequalities on marginalized communities. These communities often bear the brunt of environmental hazards, such as pollution, climate change, and natural disasters, which exacerbate their existing vulnerabilities. The economic collapse in marginalized regions due to environmental changes is a multifaceted issue with severe consequences for communities that are already vulnerable (Islam 2021). Research indicates that marginalized communities are disproportionately exposed to environmental risks, leading to social inequalities and health issues (Elsner et al., 2025; Williams et al. 2010). According to UNHCR, natural catastrophes are now three times more likely to cause displacement than conflict and violence (*Flüchtlingshilfe, Fluchtursache Klimawandel. Klimaflüchtlinge Weltweit* 2024). At the same time, they are frequently excluded from decision-making processes and lack access to resources that could mitigate these impacts. Transformative actions are essential to address these injustices, ensuring equitable distribution of environmental benefits and burdens, integrating the voices and needs of those most affected by socio-environmental challenges, and empowering marginalized groups through inclusive policies and practices.

Jesuit Worldwide Learning understands marginalization in terms of access to higher education as a profound and multifaceted issue impacting various excluded communities. These include forcibly displaced individuals, particularly the youth, who are refugees or internally displaced, residing in camps or urban areas, as well as returnees and those affected by conflicts. Worldwide, only 7% of refugee youth have access to higher education compared to the global average of around 40% (UNESCO 2022; UNHCR 2023). Additionally, JWL recognizes the plight of oppressed communities living under autocratic regimes where ethnic, social, cultural, religious, and gender discrimination curtail their freedoms and human rights. For Indigenous communities, marginalization is experienced in many ways, particularly due to (neo) colonialism, economic exploitation, injustice, and systemic and direct violence. Furthermore, poor communities, whether in rural or urban settings, face economic violence that systematically limits their access to higher education. Furthermore, several Indigenous researchers point out that Indigenous communities are affected by the consequences of climate change without being involved in the root causes and see this as a continuation of colonialism (Moore 2016; Young 2021). Moore (2016) suggested the term "Capitalocene" to describe these dynamics.

In this light, ecological transformation and peace are ever more needed for people at the margins. Regardless of the context of marginalization, the story is often similar. Marginalized people all face in various ways the effects of climate change and different forms of environmental and social injustice. This call for a better future resonates with the concept of Active Hope, developed by Macy and Johnstone (2012), which, in its name, urges action. This concept can serve as a framework to better understand the praxis of JWL through the BASD program as one way to "tackle the mess we're in" (Macy And Johnstone 2012, p. 19) and contribute to the discussion on ecological transformation for peace and justice.

12.2.1 The Concept of Active Hope

Active Hope is a practice that empowers individuals to become agents of change amid the challenges and uncertainties of contemporary times. As described by Macy and Johnstone (2012), Active Hope involves three key steps: taking a clear view of reality, identifying the direction or values we hope to see expressed, and taking steps to move ourselves or our situation in that direction. Unlike passive hope, which depends on external outcomes, Active Hope focuses on participation and action: "Active Hope is a practice. Like tai chi or gardening. It is something we do rather than have. It is a process we can apply to any situation" (Macy And Johnstone 2012, p. 21). In the face of significant current challenges, it enables a "Third Way" (or "Great Turning") between doing Business as Usual and being paralyzed by the bleak outlook of the world. Central to this transformation is an exceptional understanding of the term "hope," that is not focused on the likelihood and the possibility of certain results but rather puts a focus on the aspect of hope as a "desire" for something better (p. 20). In that sense, "Active Hope is about becoming active participants in bringing about what we hope for" (p. 20).

This Great Turning has three dimensions. Firstly, it is important to "hold actions," to raise awareness and document the damage (p. 41). This is essential, but insufficient. Secondly, it needs "Life-Sustaining Systems and Practices," a "creative re-design" of our practices to reduce harm and introduce sustainable solutions (p. 42). However, this would still be incomplete without the third dimension, a "Shift in Consciousness" (p.43) with "Compassion and Caring" at its core (p.43). This compassion for others established a new understanding of connectedness in us, giving the name for Macy's work, "The Work that Reconnects" (p. 49). This work "maps out an empowerment process that journeys through four successive movements, or stations, described as Coming from Gratitude, Honoring Our Pain for the World, Seeing with New Eyes, and Going Forth" (ibid). Whereas showing gratitude is a well-known way to "promote a sense of wellbeing" and helps us to "build trust and generosity" (p. 54–66), the idea of Honoring our Pain emphasizes that only factual information is not enough. It rather needs a compassionate approach toward our crises that allows us to "digest" that information and share our various emotions (p. 67–88). These first two steps enable us to develop a "wider sense of self" (p. 90), where our "personal identity becomes rooted in something larger" than ourselves (p. 93). It also introduces us to a new understanding of power (p. 107), that is not competitive and exploitative, but collaborative and empowering and ultimately helps us to have a "richer experience of Community" (p. 121). All of this combined will enable and motivate us to act by "identifying a destination that could be reached, a reality that could be created" (p. 156).

In all aspects of this concept of Active Hope, the authors continuously stress the importance of community, whether it is on a personal level or considering the earth's community. By working together, we can achieve more than we could individually, which they call emergence (p. 113), and our networks support us, be it through concepts like "Co-Intelligence" (p. 167) or so-called "Study-Action-Groups" (p. 195).

The concept of Active Hope offers an alternative framework to view hope in practice and understand its role in the context of higher education at the margins. Rooted in compassion and collective empowerment, Active Hope emphasizes the power of community and interconnectedness, two areas that deserve to be investigated when assessing the impact of higher education on students at the margins.

12.2.2 Purpose of Higher Education and Hope

This concept of Active Hope calls to attention what has been called the "third mission" of universities. The third mission of HEIs is often considered to be the non-academic mission, encompassing social responsibility, community engagement, knowledge transfer and dissemination, advocacy and policymaking (Berghaeuser And Hoelscher 2020; Pausits 2015). Following the first and second mission of HEIs, which are linked respectively to education and scientific research, the third mission calls on universities to be agent of change in the society they are embedded in and act as problem-solvers of the challenges of their current times.

Scholars have long argued that HEIs should hold a transformative role in society by empowering marginalized communities and fostering broader social impact (Barnett 2021, Freire 1970, Santos 2006). For instance, Barnett (2021) proposes a model of higher education rooted in ecology, advocating for universities to adopt an ethic of critical stewardship. His concept of the "Ecocene" envisions a planetary era in which humanity and nature coexist in harmony, with universities redefined as key actors in addressing the Anthropocene's challenges. This reimagining of higher education underscores its capacity to shape sustainable futures and act upon the hope of a peaceful and environmentally conscious world.

This ultimately highlights the role of hope as a driver for HEIs to holistically realize their mission's role in addressing these issues and building a more sustainable world (Cortese 2003). For Paulo Freire, the father of critical pedagogy, "without hope, there is no way we can start thinking about education" (Freire 2007, p. 87, as cited in Lorio And Tanabe 2019). Education, in this sense, becomes a pursuit of hope, where individuals discover their potential and transcend the limitations of fatalistic narratives. HEIs not only prepare future generations to tackle global challenges, but also create spaces for transformative learning, growth, and the pursuit of dreams, transcending the fatalistic narrative of neoliberalism (Lorio And Tanabe 2019).

This vision of an ecological university aligns with the concept of Active Hope, as HEIs can embrace their responsibility to foster sustainability, community engagement, and planetary stewardship. Active Hope calls for transformative action, resilience, and collective effort in addressing global challenges—principles that resonate deeply with the ecological university's commitment to harmony between humanity and nature. At JWL, programs like the Bachelor of Arts in Sustainable Development (BASD) embody this ethos, equipping students with the tools to become agents of change in their communities. In the following section, we investigate, through the case of the BASD program, how higher education can inspire both hope and action for a sustainable future.

12.3 The Bachelor in Sustainable Development: A Learning Space for Ecological Peace Leaders

To realize the third mission of HEIs and foster hope and socio-environmental justice, the institutions Xavier Institute of Management (XIM) University in Bhubaneswar, India, and Jesuit Worldwide Learning have joined forces to initiate a blended learning Bachelor of Arts in Sustainable Development for communities at the margins. XIM University in Bhubaneswar is committed to providing "education for the margins" by making education more inclusive and affordable. JWL, with almost 15 years of experience, has developed a model based on scientific knowledge and research that enables it to roll out higher education programs globally for students in marginalized regions, such as refugee camps, conflict settings, rural areas, and Indigenous land (Mayr And Oppl 2022). Following their order's universal apostolic preferences and the (similar) ideals of Pope Francis's encyclical letter Laudato si' (Francis 2015), JWL, as a Jesuit organization, also has an intrinsic motivation to provide transformative digital higher education for marginalized target groups in the field of sustainable development.

In this context, we propose understanding the role of higher education in nurturing Active Hope and promoting ecological peace at the margins through a case study of the BASD program. This analysis will examine its pedagogical model, course design, curriculum, and a specific course, the Peace Leader.

12.3.1 Blended-Learning Fostering a Practice of Active Hope in Higher Education for Marginalized Communities

Over the past decade, technological advancements have created new opportunities for learners in marginalized settings. As Barnett (2021) notes, "The university and its activities are in fundamentally new time-space configurations in which physical place seems to matter little" (p. 190). The emergence of numerous online and blended learning initiatives for refugee higher education in the early 2010s was facilitated by new technological developments in online education (Abuwandi 2019; Burde et al. 2017). Refugees, frequently residing in marginalized areas, such as camps with limited educational infrastructure, have restricted mobility, often bound to specific rules of freedom of movement, which can disrupt their academic pursuits (Dryden-Peterson et al. 2017). Online learning helps overcome these physical challenges by addressing the shortage of qualified teaching staff and the inadequate higher education infrastructure in refugee environments (Giles And Miller 2021). As a result, it meets the global demand for accessible higher education for refugees, connecting them with professors worldwide and bridging the educational gap.

Nevertheless, while digital education is seen as a beneficial tool to enhance access to higher education for students in Indigenous communities (Brant 2023; Reedy 2019; Tessaro And Restoule 2022; Wilks et al. 2017), there is a consensus through the literature of online learning for learners in emergencies, that local support is crucial for learner success (Giles And Miller 2021; Gladwell et al. 2016; Halkic and Arnold 2019; Mayr and Oppl 2023). The characteristics of facilitators are particularly important for ensuring student success in low-resource contexts (Mayr and Oppl 2023). This local aspect of blended learning also helps build a learning community, which is vital for developing intercultural and interpersonal skills, especially in refugee camps (Baker et al. 2018; Rega And Honen-Delmar 2022). Additionally, blended learning programs with reduced scheduled presence time are an approach to overcome time conflicts between employment and education, by meeting students' needs with increased flexibility in study time. These characteristics of blended learning demonstrate how blended learning can be beneficial for learners living in contexts of marginalization (Abuwandi 2019; Giles And Miller 2021; Gladwell et al. 2016).

The bachelor's program in Sustainable Development at XIM University in Bhubaneswar, India, newly designed and delivered by JWL in 2020, uses a blended learning model with two key roles: onsite and online facilitators. Onsite facilitators—often local graduates—support student groups, encourage community learning, and help reflect on course content in a cultural context, embodying the Jesuit value of accompaniment. Online facilitators, appointed by the university, handle grading, feedback, and academic support.

These elements are essential to JWL's delivery model and students' success in the BASD program as they provide the needed accompaniment and servant leadership—a leadership style driven by a willingness of serving others first (Greenleaf 2005). Hence, this flexible and accessible model enables to reach anyone, regardless of how remote they are from a university, as blended learning connects HEIs to marginalized communities and enables knowledge to be accessible to future ecopeace leaders around the world. While the blended learning model sets the frame for the use and role of technology in the pedagogical approach, the course design which will be elaborated in the following chapter, provides more details about the dynamics and structures of the learning process.

12.3.2 Course Design for Transformative Learning

JWL's blended learning programs consist of 8-week courses, each containing 8 units with defined learning objectives. The dynamic of JWL's academic delivery model follows a pedagogical approach similar to Kolb's experiential learning cycle (Kolb 1984) and employs organizational scaffolds (Van De Pol et al. 2010) to structure learning activities in challenging environments. The core elements—Context, Experience, Reflection, Action, and Evaluation—recur in each unit of a JWL course and are based on the Ignatian Paradigm elaborated in the practical guide of Ignatian pedagogy (Duminuco n.d.).

As visualized in the following graphic (Marek And Walulik 2022), understanding the context that frames the learning circumstances is crucial. Subject Matter Experts (SMEs) who develop the course content, the instructional designers, the online and onsite facilitators, and the students' living environments are all essential elements for the foundation of any learning process (2022, p. 4488) (Fig. 12.1).

The aspects of this paradigm are taken into account in the development of each JWL course by the instructional designers who assist the SMEs in the development of the learning materials. The following section lines out which aspects of the course design incorporate the Ignatian paradigm aspects of context, experience, action, reflection, and evaluation. The principle of locally rooted and culturally based learning experiences is embedded in JWL instructional design through the central aspect of context. In alignment with JWL's vision to educate men and women for others, leading to genuine transformation, all courses are intentionally designed to integrate learners' cultural and community realities into the educational experience. JWL's learning materials include photos, case studies from various JWL learning centers, and structured learning activities that encourage learners to engage with their

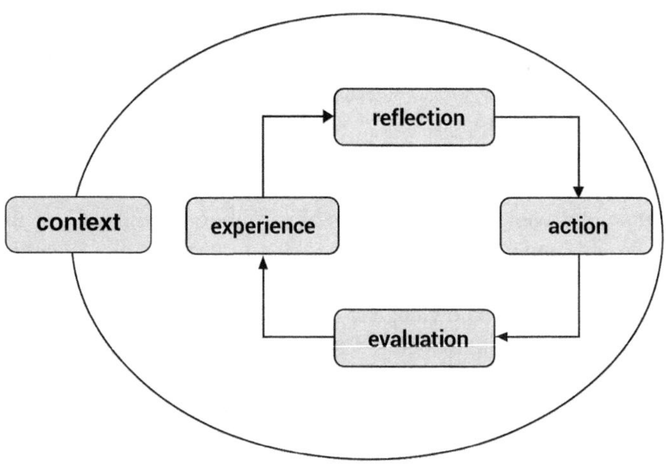

Fig. 12.1 Diagrammatic interpretation of Ignatian education (based on Society of Jesus) (1993) (Marek And Walulik 2022, p. 4488)

immediate environments. These activities regularly involve exploring local issues, conducting interviews with community members, field observations, and sharing personal and regional insights in a global virtual classroom. This exchange not only promotes intercultural dialog, but also deepens learners' reflections on global concepts in light of their diverse local realities. Each course culminates in a project-based assignment that requires the application of course knowledge in the learner's local context. Through this model, JWL ensures that learning is never abstract but is instead deeply connected to live experience, fostering both local relevance and global understanding.

According to the blended learning concept, students engage in onsite learning groups to apply their learnings in actions or reflections in their community. The course design incorporates the aspect *experience* of the Ignatian paradigm in various ways. During onsite meetings, activities for team building, group work, and community engagement are assigned. These tasks help contextualize the content, facilitate face-to-face interaction, and provide necessary support and social inclusion in collaborative learning processes. The Global Classroom, where learners interact and submit their weekly online work, is overseen by a professional faculty member to ensure academic quality in student support and grading. Individual learning experiences can also be shared and discussed in both global and onsite classes. The content and assignments are developed specifically for students in marginalized regions, in collaboration with subject matter experts from accrediting university partners and the evaluations and feedback of onsite facilitators are implemented in course revision to create meaningful learning experiences for students.

Regular *reflection* activities are part of each course to help students clarify their learning process and transfer the acquired knowledge into practice. The onsite facilitators are trained to support students in learning with reflection impulses to create a mindset where reflective impulses lead to actions and experiences.

Active learning, where students are involved in hands-on activities to apply their knowledge, is also encouraged in JWL's Course design to facilitate the transfer of learning. Each weekly final assignment builds on previous content, reinforcing learning through ongoing application. Every course includes a project work that applies course concepts in the student's own context, supported by online facilitator feedback and reflection. Onsite facilitators help organize and implement these projects within local communities.

Evaluation is an essential component of this learning process. Constructive feedback, qualitative assessments, and continuous student support are integrated to enhance motivation, self-perception, and individual growth. Self-check quizzes included in the self-study materials help reinforce learning, while onsite facilitators provide qualitative feedback and support during face-to-face exchanges. Online facilitators contribute further by offering detailed feedback on weekly assessments, ensuring that students receive continuous support throughout their learning journey (Mayr And Oppl 2022).

In addition to considering the aspects of the Ignatian Pedagogy Paradigm, the course design principles applied in the BASD program follow the framework of Transformative Sustainability Learning (TSL) (Sipos et al. 2008). Following this framework, the course design applies the aspects of head, heart, and hands in the learning process. Repetition and diverse activities to get involved actively in the learning process deepen students' understanding of the subject matter to acquire cognitive (head) learning objectives. Frequent reflections help students maintain the connection between their learning and their personal development. Action-focused activities, combined with reflective approaches, nurture students' personal growth and the affective domain of learning (heart), while hands-on activities such as field-work and project work enable them to apply what they have learned in practical settings (hands), the so-called psychomotor domain of learning (Sipos et al. 2008). According to the mission and vision of JWL, this holistic approach—shaping the students' "heads," "hearts," and "hands"—seeks to foster the development of individuals who are committed to serving others. Through education, students can grow as men and women for others, transforming their communities by learning together and applying their knowledge to benefit their communities (Balleis et al. 2020; Rega And Honen-Delmar 2022). This shift forms Self-Centeredness to Other-Centeredness is understood by Macy and Johnson as a "Shift in Consciousness" with "Compassion and Caring" at its fundament (Macy And Johnstone 2012, p. 43). This is the third dimension of the Great Turning as Macy and Johnstone labeled their concept of the transformation in the mindset of understanding hope toward "Active Hope is about becoming active participants in bringing about what we hope for" (Macy And Johnstone 2012, p. 20).

As each course of the bachelor program includes a project work, they are designed in such a way that by week 4 at the latest, enough basic concepts are explained to introduce a practical task and to guide students step by step toward a small project. The small project involves developing an idea for a short project in the subject area of the course. For instance, in the class on waste management, one of these small projects involves changing existing practices or introducing new processes to create more sustainability in their living environment. After completing these small practical project designs in each course, there is a final Project Work at the end of the program, which focuses on the concept and implementation of such a practical project. The didactic design element of the project assignment in each course, as well as the final project, contributes to the transformation of the understanding of hope. As Macy and Johnson (2012) describe, the second dimension to strengthen Active Hope understanding is related to rethinking and creatively adapting existing systems and practices for a more sustainable life, which is encapsulated in the practical projects throughout the BASD program.

12.3.3 Curriculum

While the BASD blended learning and course design are essential factors for creative ecological transformation at the margins and fostering active hope among young leaders, the curriculum is key in understanding how these students are equipped with knowledge and skills to contribute to this ecological peace development. The

BASD program is indeed thematically aligned with the United Nations' Sustainable Development Goals (SDGs). For all fields of action identified in the framework, the program offers courses that introduce students to the basic concepts and professional fields of action in these areas. There are 3 focus areas: Sustainability, Leadership for Peace Building, and Business Enterprise & Management. The program learning objectives are listed in the following table and address all three thematic areas.

As stated by Lampoltshammer et al., "The SDGs are defined in a way that they are not silos but encourage synergies between one or several goals" (2021, p. 3). This dynamic is also pursued in the course syllabus, as most courses are designed to promote mutual understanding in one SDG and also foster recognition of the interconnectedness with other SDGs. Additionally, teaching in the field of Sustainability is not only about subject knowledge in sustainability, but also about developing students' sustainability attitudes and behaviors. Therefore, this sustainable development pedagogical approach seeks to center itself on the individual transformation as a starting point toward a sustainable future (Braßler And Sprenger 2021, p. 2).

A closer look at the various courses available in the BASD program helps to understand how this program nurtures the individual's views about their environment, while seeking to form men and women for others. The bachelor's program comprises a total of 148.5 credits, which XIM University in Bhubaneswar, India, accredits. The program is composed of 24 courses, which have been developed by professors of different departments of XIM University, such as "Climate Change," "Fundamentals of Environmental Science," "Waste Management for Sustainable Livelihood," "Sustainable Energy Systems," and "Environment and Social Impact Assessment." In addition, 5 courses from the Newmann Institute in Sweden are part of the program, such as "Humans and Environment," "Agroforestry," or "Responding to Environmental Challenges." In addition, the Peace Leader courses are part of the BASD program with 3 modules on the subjects of self-care and conflict coaching, conflict theories, and peace studies. These courses are accredited at the Hekima University College Institute of Peace Studies and International Relations.

While all the BASD courses contribute to building knowledge and skills in the sustainability field, a closer look on one course of the BASD program, the Peace Leader course, can help us to understand the impact of higher education programs on graduates and their communities and how these courses nurture Active Hope and ecological Peace Leaders.

12.4 The Peace Leader Program: Nurturing Active Hope and Servant Leaders

The Peace Leader program, while being a mandatory course of the BASD program, has also been offered as a standalone program, also accredited by Hekima University College since 2019 to students at the margins around the world. As the BASD program seeks to form holistic leaders in sustainable development, the topic of Peace

Leadership became an evident option for XIM University and JWL while developing the curriculum of the bachelor program.

Like the rest of the BASD courses, the Peace Leader program utilizes a blended learning approach. Course materials are accessible through the JWL Learning Management System (known as the Humanitarian eLearning Platform—HelP) or the JWL Global E-learning app, which allows for offline access to content. Through the various assignments, students reflect on their experiences and apply their learning through five project works during the program. These projects involve activities that engage local leaders and communities such as interviews or shadowing, encouraging students to apply their knowledge within their own contexts and contribute to their communities.

By the end of 2023, the Peace Leader program had been delivered nine times, reaching a total of 639 participants across 36 community learning centers in 11 different countries. Currently, 400 young people have successfully completed the program. In 2022, a study was conducted to research the impact of the Peace Leader program on the lives of graduates and their communities. After analyzing the voices of 30 graduates who completed the program before September 2021, a conceptual model was developed demonstrating the impact of the Peace Leader course on graduates and their communities (Futuro et al. 2023). Among the 30 graduates, 4 lived in Kakuma refugee camp, a semi-arid area in northern Kenya that has been home to 290,000 refugees—primarily from South Sudan, Sudan, and Somalia—since 1992 (UNHCR 2024), while 8 were located in remote regions of Afghanistan. Another 3 graduates were based in Dzaleka Refugee Camp, situated 40 km outside Lilongwe, the capital of Malawi. This camp accommodates approximately 50,000 refugees from neighboring countries, including the Democratic Republic of Congo, Burundi, and Rwanda (*Malawi*, n.d.). Additionally, 4 graduates came from a marginalized, low-caste fishing community in Sneharam, Kerala, India. The remaining 8 were based in Kurdistan, Iraq, either as Syrian refugees living in camps or the capital city, or as Yazidis who had been forcibly displaced due to conflict and genocide in the 2000s.

The impact model of the Peace Leader course is briefly presented in light of understanding how Peace Leader graduates, through their actions, actualize the concept of Active Hope in their communities (Fig. 12.2).

After a methodic thematic coding analysis (Braun And Clarke 2012), the following three main dimensions of the impact emerged. First, the Peace Leader course has been seen as developing graduates' self-control, self-confidence, and self-awareness, which triggers a personal transformation. For instance, this graduate in a small fisherman underprivileged community in Kerala, India highlights the significance of introspection emphasized in the Peace Leader course and explains how specific practices learned during the course assist in maintaining their journey of self-awareness:

"Everything we are taught like that—who am I? What do I live with? [...] And another thing, confidence level, is whenever decrease the confidence, [...] I practice the four breathing, presenting practice a lot. Like I already said, I am not religious, I am spiritual, it has helped me a lot."

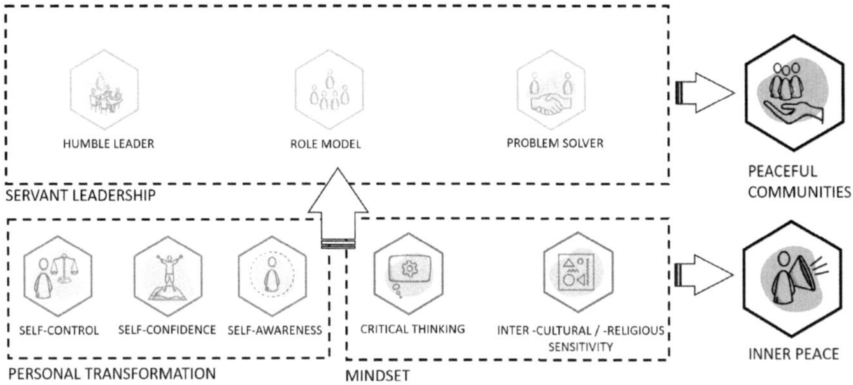

Fig. 12.2 The conceptual model (Futuro et al. 2023, p. 8)

This spiritual aspect is also essential for Macy and Johnstone, as it helps us to build trust and become more resilient (Macey And Johnstone 2012).

Furthermore, this focus on self-prioritization and the journey to uncover and embrace hidden aspects of oneself fosters personal transformation, according to graduates, notably in Kerala, India and Domiz refugee camps in Iraq. Graduates acknowledge the intricate emotions that this journey evokes as they grow in self-awareness. A graduate in Kerala, India, for instance explains:

"In each unit, the problems were arising inside of me because, it's peeling my inner self. It's feeling very difficult because every time I think like this […] it's disturbing. […] We have a pickle, a bitter experience in our life and every time we take the bottle open and take the little bit, take the taste, then closed it. […] So, after doing this course, I collapsed this bigger bottle, and I'm very free and I am taking decisions and saying something to others. I cannot judge anything for I am doing my own work. Nothing affects me. Because I have a character […]. So, how to handle our feelings? Peace Leader is mostly helped like that. […] so when a conflict has happened, our energy model will come, so I understand that, and I am dealing with that inner model, now I am peacefully and smiley."

Graduates frequently cite self-control, closely linked to self-awareness, as a crucial factor in managing inner conflict. This graduate in Khanke Internally Displaced Person (IDP) camp in Iraq elaborates:

"the inner conflicts, with something very bad actually, so it taught me how to deal with the inner conflict and, just like to transfer from you know conflict to peace, internally. It was something great for me because, like now, […] I can control my feelings, I can manage how to deal with conflicts, whether it's internal or external."

Second, graduates share their perspectives about how the Peace Leader course impacted their mindset, adopting critical thinking and interreligious/intercultural competence skills. For instance, this graduate living in the refugee camp of Domiz, Iraq, explained how the course helped them to use a new kind of thinking to improve their daily surroundings:

"the results in my daily life issues are a great witness and proof for my critical thinking improvement. I better control the situation with the possible outcomes that comes from my thinking and people around me have also started trusting this quality that has freshly came into me."

This transformation in the students' mindset is also reflected in their views toward diversity, opening new realms for interreligious/intercultural sensitivity skills. One compelling example of such transformation is provided by a graduate living in Erbil, the capital city of Kurdistan, Iraq. He notes that the course significantly influenced his intercultural and interreligious perspective through his interaction with his onsite facilitator, a Muslim with a different religious background. The graduate explains that, due to the persecution of the Yazidi community, he had previously struggled to engage with Muslims. However, his interaction with the facilitator during the course revealed to him that the facilitator was "a great person," demonstrating his shift in perspective:

"this course helps all of us, just like to feel open-minded and to talk about things like this […] Some people ask us this, or some such questions so what is it like having, you know, different people […] from different religions and cultures […] and, of course, the course helped a lot, and because our own onsite facilitator was Muslim."

Another graduate mentioned the ability to interact with others more positively, which an Iraqi graduate described as "outer peace." They further elaborated on this concept as this IDP graduate living in Khanke, Iraq states:

"… being good, understanding all we are doing together […], accepting others with their differences. All this can be called peace, the real peace, not the one that we said only if there's no fight, then there is peace."

Therefore, this course enabled students to realize the complexity of achieving peace by engaging with others, learning collectively and inviting them to rethink prejudices. This ultimately feeds back into this transformative process of achieving inner peace.

Third, graduates reflect a growth in their leadership, adopting servant leadership attributes, such as humility, role-modeling, and problem-solving skills, which is seen as developing peace around their communities. For graduates, the realm of leadership, along with the development of self-confidence and a sense of belonging to the community, can also be linked to their sense of responsibility toward others in the community. For instance, a graduate in Domiz refugee camp, Iraq asserts:

"Earlier when I would walk around the street, I think I wouldn't look around me I would just look straight on the ground and reach to wherever I'm going, my destination […] Even in my house, my home, when I would hear conflicts and I think these conflicts are all around us. I would just ignore them because I would say none of my business. […]. But after this course I think […] it made me realize that this community is a part of my life. And I'm a big part of it. […] And in order to leave your […] footprint on it, you need to try to help them and be an active member of this community. So always get involved always […], don't just make boundaries around yourself between yourself and your community because this community leaves a huge impact on us and it forms our future selves in many ways."

This view toward hope and community aligns with the concept of Active Hope in that a "rich Experience of Community" is essential in the journey of developing a wider sense of self which will empower us to move on and engage (Macy And Johnstone 2012).

Many graduates demonstrated how they enacted their hopes and turned them into benefits for the community, reflecting their servant leadership attributes. For example, a 25-year-old graduate, who identifies as an IDP residing in Khanke camp, Iraq, offered mental health support to Yazidi community members. Around 2014, the Yazidis were persecuted by the Islamic State, leading to the forced displacement of many Yazidis. He explains that he works with "survivors from the hands of ISIS. So, I facilitate, you know all the meetings and trainings they have, and I also interpret for them, translate all the documents, stories they have." He taught them the four breath-presencing techniques, skills he had acquired during the course that helped him manage his emotions. In addition to offering this support, at the time of the interview, he was in the process of establishing his own teaching center for survivors. He explains:

"One of the most important points is just like teaching survivors, since it's been years and years. Of course, after the genocide, I know thousands of survivors were, you know, prevented from studying and maybe making their dreams true. So, I said, at least we can just teach them, you know language, IT skills, maybe provide some MHPSS [Mental Health and Psychosocial Support] or something."

These examples illustrate the power of graduates to embody the concept of Active Hope in their communities, appearing as Peace Leaders, ultimately creating a more peaceful relationship with themselves and their communities. These Peace Leader graduates' testimonies are only a selected few among the hundreds of students who took this course, either as a standalone program or as part of the BASD program. To capture the insights from the graduates who are also BASD students, a brief analysis of internship reports of BASD students was conducted to demonstrate further how Peace Leader graduates in their BASD studies become servant leaders promoting peaceful environments, conscious of sustainability needs in their community. One BASD student in Erbil, Iraq, who is also a graduate of the Peace Leader program, contributed to making her environment more sustainable by analyzing Energy Efficiency and Waste Management Processes during her internship in the hospitality sector. The Experience of her achievements has also a long-lasting empowering effect, helping us to "keep a hold on our vision" (Macy And Johnstone 2012, p. 117) as she states in her internship report:

"I gained practical insights into the challenges and opportunities of integrating sustainable practices in a business setting. The experience has motivated me to pursue a career where I can actively contribute to sustainable development initiatives within the corporate sector, aligning my professional journey with my life goal of making a positive impact on the world."

Another graduate in Dzaleka refugee camp (Malawi), who is currently a BASD student and the Peace Leader onsite facilitator, did an internship at Poverty Alleviation Program to explore activities that effectively localize SDGs, with a specific focus on poverty eradication in Malawi. Her research showed that despite challenges, such as corruption or economic pressures, the initiative has been productive, helping to reduce hunger and poverty and that effective management skills are crucial for beneficiaries to maximize the program's benefits. She shares in her internship report that:

"This internship was truly a great opportunity for me not only to learn how the organization working on sustainable initiatives run, but also to open my eyes to understanding how we can apply what we learn in our community to bring great impact on the environment, society, and people."

These examples illustrate how the BASD program and, particularly through the Peace Leader courses, enable men and women to build peace with themselves, their communities, and their environment at large. Therefore, The Peace Leader program stands as a transformative example of how higher education, particularly within the BASD curriculum, can foster active hope and drive ecological and social change. By empowering students with self-awareness, servant leadership attributes, and a commitment to sustainability, the program equips graduates to embody peace leadership in their personal lives and within their communities. Whether through addressing local conflicts, providing mental health support, or advancing sustainable practices, Peace Leader graduates illustrate the profound effect of education in creating resilient, hope-driven individuals capable of promoting ecological justice and fostering peaceful, sustainable environments.

12.5 Conclusion

In an era marked by ecological crises and deepening social injustices, higher education has a unique opportunity—and responsibility—to foster hope and empower marginalized communities to lead transformative change. Drawing on the concept of Active Hope (Macy And Johnstone 2012), this paper has illustrated how hope must be actionable and purpose-driven, particularly in contexts of displacement, oppression, and environmental degradation. The JWL BASD program exemplifies how higher education can not only reach the margins, but also empower individuals to become servant leaders who act for the betterment of their communities and the care of our typical home.

By integrating transformative learning approaches, such as Ignatian pedagogy, and emphasizing sustainability, leadership for peacebuilding, and enterprise management, the BASD program reflects a holistic model of education. This design fosters critical thinking, self-awareness, and leadership skills while equipping learners with the tools to address interconnected environmental and social challenges. It aligns with Barnett (2021)'s vision of the ecological university, where the interconnectedness of humanity and the planet is recognized, and higher education becomes central to addressing global crises.

The impact of such programs is not merely academic or individual. As seen through the transformative outcomes of BASD and Peace Leader graduates—enhanced confidence, critical mindsets, and servant leadership—higher education at the margins becomes a catalyst for ecological peace. This peace extends beyond personal growth to foster community resilience, social harmony, and collective action. For marginalized learners, education is not simply an opportunity but a life-saving necessity, offering a pathway out of systemic injustice and a means to build a more just, inclusive, and sustainable world.

Higher education institutions (HEI) must therefore fulfill their third mission with a renewed sense of purpose—ensuring that their role in solving global challenges is holistic, equitable, and inclusive. Through programs like the BASD, HEIs can contribute to the integration of marginalized voices into policies and programs to build sustainable futures. In this light, higher education can embody Active Hope, nurturing leaders who are transformed and who, in turn, transform their communities. By prioritizing ecological peace and justice, HEIs can contribute meaningfully to the ecocene, a world where equality, inclusion, and interconnection are the foundation for collective flourishing.

Therefore, the contribution of this paper to the literature lies in its emphasis on bringing the voices of marginalized communities to the center of conversations on ecological justice and peace—an essential perspective when considering solutions to global issues. This work highlights how the framework of Active Hope resonates deeply with those at the margins, where hope is not merely aspirational but a critical and life-saving force. In contexts marked by war, displacement, corruption, and systemic injustice, imagining hope can feel impossible. Yet, this study reveals that hope when rooted in action becomes a transformative tool, not just for personal resilience but for empowering others and fostering collective change. Higher education emerges as a key vector in this process, equipping individuals with the knowledge and agency to rebuild their lives and their communities. This is a call to HEIs to fully embrace their role in contributing to social and ecological justice, ensuring that active hope becomes both a reality and a driving force in shaping a sustainable and peaceful world.

While this paper has primarily focused on the Peace Leader courses within the BASD program, further research is needed to assess the broader impact of the BASD curriculum across all courses and learning pathways. Such exploration would offer a more comprehensive understanding of how holistic education contributes to sustainable development. At the same time, it is important to recognize that higher education alone cannot achieve ecological peace and justice. Future research should investigate how higher education can be effectively integrated with social, financial, and political interventions to create more systemic and lasting change. Finally, a deeper inquiry into how HEIs understand and operationalize their third mission, especially in relation to sustainability and equity, is essential for evaluating their contribution to more inclusive and transformative futures. Ultimately, addressing the complexity of the world's ecological and social challenges requires interdisciplinary research and the inclusion of diverse voices, particularly those from the margins, to truly pave the way toward ecological peace.

Acknowledgments Special thanks go to Dr. Isabella Rega, JWL former Research Director, for her leadership, guidance, and expertise throughout the impact study of the Peace Leader program.

Appreciation is also extended to the dedicated research assistants: Carolina Futuro in Geneva, who diligently analyzed the interviews and drafted the Peace Leader report, and the local research assistants, Ghak Atem, Patricia Kazembe, Noah Lapp, Farzana Farzam, Zainab Hamidi, Fr. Benny Ouso, Jacob Mohamed Ali, whose efforts in data collection made this study possible. Sincere gratitude is also extended to Dr. Mary McFarland and Lassalle Institute readers who provided valuable feedback, enriching the quality and depth of this chapter.

References

Abuwandi S (2019) Blended higher education opportunities for refugees: a comparative study

Acosta Güemes L, Cusumano AM, Giménez Milán C (2023) Institutions of higher education from the beginnings of civilization to 600 AD. Asian J Educ Soc Stud 46(3):29–40

An Z (2021) The higher education paradigm shift. Institutions (HEIs) 4(1):35–42

Baker S, Ramsay G, Irwin E, Miles L (2018) 'Hot', 'Cold' and 'Warm' supports: towards theorising where refugee students go for assistance at university. Teach High Educ 23(1):1–16. https://doi.org/10.1080/13562517.2017.1332028

Balleis P, Ablewhite S, Bathany C, Borja A, Carminati A, Hengst S, Mayer P, Mayr A, Rega I, Samusarma S (2020) Growing Global Higher Education at the Margins—understanding JWL's blended mobile learning model. [Public-wbt, SCORM]. Jesuit Worldwide Learning - Higher Education at the Margins. https://learning.jwl.global/en/public-wbt

Barnett R (2021) The philosophy of higher education: a critical introduction, 1st edn. Routledge. https://doi.org/10.4324/9781003102939

Berghaeuser H, Hoelscher M (2020) Reinventing the third mission of higher education in Germany: political frameworks and universities' reactions. Tert Educ Manag 26(1):57–76. https://doi.org/10.1007/s11233-019-09030-3

Brant J (2023) Lessons from our Sweetgrass baskets: a Wholistic vision of academic success for indigenous women in higher education. Can J High Educ 52(4):27–40. https://doi.org/10.47678/cjhe.vi0.189759

Braßler M, Sprenger S (2021) Fostering sustainability knowledge, attitudes, and behaviours through a tutor-supported interdisciplinary course in education for sustainable development. Sustainability 13(6):3494. https://doi.org/10.3390/su13063494

Braun V, Clarke V (2012) Thematic analysis. In: Cooper H, Camic PM, Long DL, Panter AT, Rindskopf D, Sher KJ (eds) APA handbook of research methods in psychology, vol 2: research designs: quantitative, qualitative, neuropsychological, and biological. American Psychological Association, pp 57–71. https://doi.org/10.1037/13620-004

Burde D, Kapit A, Wahl RL, Guven O, Skarpeteig MI (2017) Education in emergencies: a review of theory and research. Rev Educ Res 87(3):619–658. https://doi.org/10.3102/0034654316671594

Cairns LG (2023) Socio-cultural and historical influences on 20th to 21st century internationalization of higher education. In: Global perspectives on the internationalization of higher education. IGI Global, pp 75–92

Cortese AD (2003) The critical role of higher education in creating a sustainable future. Plan High Educ 31(3):15–22

Dryden-Peterson S, Dahya N, Adelman E (2017) Pathways to educational success among refugees: connecting locally and globally situated resources. Am Educ Res J 54(6):1011–1047

Duminuco VJ, SJ (n.d.) Ignatian pedagogy: a practical approach. Society of Jesus. https://www.sjweb.info/documents/education/pedagogy_en.pdf

Elsner M, Atkinson G, Saadia Z (2025) Global risks report. World Economic Forum. https://www.weforum.org/publications/global-risks-report-2025/

Flüchtlingshilfe, Fluchtursache Klimawandel. Klimaflüchtlinge weltweit (2024). https://www.uno-fluechtlingshilfe.de/hilfe-weltweit/themen/fluchtursachen/klimawandel
Francis (2015) Encyclical on climate change & inequality: on care for our common home. Melville House
Futuro C, Rega I, Honen-Delmar M, Balleis P (2023) Peace leader: a transformative programme for the self and the community. In: Schwager R, Walach M, Heit AM (eds) Politik des Evangeliums/politics of the gospel: festschrift für Wolfgang palaver. Innsbruck University Press, pp 407–430
Freire P (1970) Pedagogy of the oppressed (trans: Ramos MB). Continuum International, New York
Giles W, Miller L (2021) Borderless higher education for refugees: lessons from the Dadaab refugee camps: Wenona Giles. Bloomsbury Academic. https://www.bloomsbury.com/uk/borderless-higher-education-for-refugees-9781350151239/
Giroux HA (2016) Disposable futures: neoliberalism's assault on higher education. Límite Revista Interdisciplinaria de Filosofía y Psicología 11(35):7–17
Gladwell C, Hollow D, Robinson A, Norman B, Bowerman E, Mitchell J, Floremont F, Hutchinson P (2016) Higher education for refugees in low resource environments: landscape review. Jigsaw Consult, United Kingdom
Greenleaf RK (2005) Greenleaf on servant-leadership: who is the servant-leader? Int J Servant-LeadShip 1(1):19. https://doi.org/10.33972/ijsl.319
Halkic B, Arnold P (2019) Refugees and online education: student perspectives on need and support in the context of (online) higher education. Learn, Media Technol 44(3):345–364. https://doi.org/10.1080/17439884.2019.1640739
Islam MR (2021) The adverse impact of ecosystem degradation and poor governance on marginalized people. N AmN Acad Res (NAAR) J 4(3). https://doi.org/10.5281/zenodo.4644052
Kolb DA (1984) Experiential learning: experience as the source of learning and development. Prentice-Hall
Lampoltshammer TJ, Albrecht V, Raith C (2021) Teaching digital sustainability in higher education from a transdisciplinary perspective. Sustainability 13(21):12039. https://doi.org/10.3390/su132112039
Lorio JM, Tanabe C (2019) Higher education and the practice of Hope. Springer Nature. https://link.springer.com/book/10.1007/978-981-13-8645-9
Lynch K (2022) Affective equality in higher education: resisting the culture of carelessness. In: Student carers in higher education. Routledge, pp 10–27
Macy J, Johnstone C (2012) Active hope: how to face the mess we're in without going crazy. New World Library
Malawi (n.d.) UNHCR. Retrieved 18 Dec 2024 from https://www.unhcr.org/countries/malawi
Marek Z, Walulik A (2022) Ignatian spirituality as inspiration for a pedagogical theory of accompaniment. J Relig Health 61(6):4481–4498. https://doi.org/10.1007/s10943-022-01628-z
Mayr A, Oppl S (2022) Higher education at the margins—success criteria for blended learning systems for marginalized communities. Educ Inf Technol 28(3):2579–2617. https://doi.org/10.1007/s10639-022-11282-3
Mayr A, Oppl S (2023) Higher education at the margins–success criteria for blended learning systems for marginalized communities. Educ Inf Technol 28(3):2579–2617. https://doi.org/10.1007/s10639-022-11282-3
Miles EW, Miles EW (2016) Higher education from antiquity to the medieval university. In: The past, present, and future of the business school, pp 9–16
Moore J (2016) Anthropocene or Capitalocene? Nature, History, and the Crisis of Capitalism, pp 1–11
Pausits A (2015) The knowledge society and diversification of higher education: from the social contract to the Mission of universities. In: Curaj A, Matei L, Pricopie R, Salmi J, Scott P (eds) The European higher education area: between critical reflections and future policies. Springer, pp 267–284. https://doi.org/10.1007/978-3-319-20877-0_18

Perkin H (2014) History of universities. In: International higher education, vol 1. Routledge, pp 169–204

Reedy AK (2019) Rethinking online learning design to enhance the experiences of indigenous higher education students. Australas J Educ Technol 35(6):132–149. https://doi.org/10.14742/ajet.5561

Rega I, Honen-Delmar M (2022) Liberal studies and servant leadership: inspiring Ignatian values at the margins. Jesuit High Educ J 11(1) https://digitalcommons.lmu.edu/jhe/vol11/iss1/4

Sipos Y, Battisti B, Grimm K (2008) Achieving transformative sustainability learning: engaging head, hands and heart. Int J Sustain High Educ 9(1):68–86. https://doi.org/10.1108/14676370810842193

Santos B de S (2006) The university in the 21st century: toward a democratic and emancipatory university. In: Apple MW, Ball SJ, Gandin LA (eds) The routledge international handbook of the sociology of education. Routledge, pp 274–282

Tessaro D, Restoule J-P (2022) Indigenous pedagogies and online learning environments: a massive open online course case study. AlterNative Int J Indig Peoples 18(1):182–191. https://doi.org/10.1177/11771801221089685

UNESCO (2022) Higher education global data report (summary). In: A contribution to the world higher education conference, 18–20 May 2022. [Working Document]

UNHCR (2023) UNHCR reports progress in refugee education; tertiary enrolment rate hits 7 per cent. UNHCR. Retrieved 18 Dec 2024 from https://www.unhcr.org/news/briefing-notes/unhcr-reports-progress-refugee-education-tertiary-enrolment-rate-hits-7-cent

UNHCR (2024) Kenya Registered refugees and asylum-seekers. Retrieved 18 Dec 2024 from https://www.unhcr.org/ke/wp-content/uploads/sites/2/2024/07/Kenya-Statistics-Package-June-2024.pdf

Van De Pol J, Volman M, Beishuizen J (2010) Scaffolding in teacher–student interaction: a decade of research. Educ Psychol Rev 22(3):271–296. https://doi.org/10.1007/s10648-010-9127-6

Wilks J, Wilson K, Kinnane S (2017) Promoting engagement and success at university through strengthening the online learning experiences of indigenous students living and studying in remote communities. In: Frawley J, Larkin S, Smith JA (eds) Indigenous pathways, transitions and participation in higher education. Springer Singapore, pp 211–233. https://doi.org/10.1007/978-981-10-4062-7_13

Williams DR, Mohammed SA, Leavell J, Collins C (2010) Race, socioeconomic status, and health: Complexities, ongoing challenges, and research opportunities. Ann New York Acad Sci 1186(1):69–101. https://doi.org/10.1111/j.1749-6632.2009.05339.x

Young JC (2021) Environmental colonialism, digital indigeneity, and the politicization of resilience. Environ Plann E: Nat Space 4(2):230–251. https://doi.org/10.1177/2514848619898098

Mélodie Honen-Delmar PhD Student in Educational Sciences at the University of Geneva and Global Director of Research and Learning at Jesuit Worldwide Learning (JWL). Her research explores the transformative role of higher education at the margins, with a particular focus on student learning experiences. Her interests lie at the intersection of higher education and international development, including educational access, learning technologies, and the use of artificial intelligence in education.

Anna Mayr PhD Candidate at the University for Continuing Education Krems, Austria. She leads learning material and technology design and development for Jesuit Worldwide Learning (JWL). Drawing on design science research, her research explores blende-learning design principles for higher education at the margins. Her interests include culturally sustaining pedagogies, epistemic justice, and higher education in Indigenous Communities and in rural, remote, and marginalized settings.

Magdalena Nauderer Country Director Iraq at Jesuit Worldwide Learning (JWL) and CEO of the JWL Förderverein e.V. As a philosopher and higher education practitioner, she is dedicated to expanding access to higher education in marginalized and crisis-affected contexts. Her work focuses on intercultural dialogue, ethics, and the transformative potential of education for displaced and excluded communities.

Part III
The Autoethnographic Space: Transdisciplinary Practices of Hope in Institutions, Projects, and Communities

Chapter 13
Science for the Good: A Plea for a More Just and Reconciliatory Transdisciplinarity

Roberto Jaramillo Bernal S.J.

Abstract The investigation and practice of hope must have its roots in the experience of impoverished people. In this plea, two important prerequisites for a just understanding of hope are presented: (1) to consciously avoid the attitude of superiority and, (2) the notion that transdisciplinarity has to follow radical ethical standards.

Keywords Hope · Impoverished people - Poverty · Inequality · Ethical standards · Public action · Transdisciplinarity

13.1 Prerequisites for Understanding Hope

What is hope? Since the moment I was invited to this conversation, I have been asking the very same question to some people in Brazil, Peru, and Colombia (the countries where I have been working, mainly with impoverished people), as well as in Malawi, Zimbabwe and Zambia, where I recently spent some time visiting Jesuit Social Centers. What is hope? Here are some of their answers.

> "Look around: it seems that Mother Earth is also being killed. The crops are dying, the river is drying, and the trees are disappearing. It is not easy to foresee the end of this situation … but we do not lose hope. Hope has to do with dreams and good desires, especially for the children; they do not deserve to live what we have already lived. Hope is the gasoline that keeps us moving and alive".
>
> Luiz Fernando de Souza, Aldea Babacal, rio Amazonas – Amazonas, Brasil

> "It seems you are looking for hope... maybe you do not have hope. But we live out of hope every day: the hope of having something to eat, the hope of arriving home quickly and safely, and the hope of finding better ways for youngsters and children to live in the future. Even the hope of celebrating life with friends and relatives every weekend, resting, talking, dancing, and eating. We continue striving. Something better should come... that's hope. " A. Kabanda, KASISI Agricultural Training Center, Lusaka, Zambia

---

R. Jaramillo Bernal S.J. (✉)
Society of Jesus, Rome, Italy
e-mail: sjesdir@sjcuria.org

T. Karcher SJ et al. (eds.), *Making Hope Possible*, Environmental Humanities: Transformation, Governance, Ethics, Law 7,
https://doi.org/10.1007/978-3-032-17719-3_13

In order to understand such answers, let me reflect on three attitudes that will determine (or even change) our perception of the relationship between the hope of impoverished people and the pertinence or relevance of our lives as people interested in sciences.

(a) There are many people who (at this very moment) are in poverty or are impoverished, but not a single human being *is* poor. In that sense, there *are* no poor people; there *is* no truly poor person. If we do not start from the solid principle that the people who we want to help with our work are worthy, we have already begun our existential quests with a sense of superiority (whether epistemological or moral or existential), and have therefore definitively missed the point. Such a "pretended imbalance" (we on the top - the poor on the bottom) will cause us to see, to perceive, to understand, and to transform reality into something it is not.

(b) The question of poverty, misery, and inequality is a political matter and a political challenge. Poverty is not a natural fact, and we should not normalize it. Relations between human beings and between societies—whether of a religious, familial, commercial, economic, or artistic nature, etc.—are always mediated by constructed, historical, intentional, and self-interested mechanisms. These relations reflect and reproduce a distribution of resources that is always in tension and conflict, and it is up to us to resolve it. With this awareness, it is crucial to address the stigma associated with poverty and to challenge the perception that people experiencing poverty are to be blamed for their condition. Instead, we must work for a collective response of justice and compassion.

    I insist not only on justice (which is the ethical imperative), but also on *compassion* in its most plain etymological sense: "feeling of sorrow or deep tenderness for one suffering or experiencing misfortune" from the Latin roots "com" (with, together) and "pati" or " pathos" (to suffer, suffering). The truth we seek in its different forms and from various perspectives (as academics, practitioners, social activists, etc.) will appear not only in comprehension but especially in compassion.

(c) Finally, let us be mindful that in speaking about poverty, violence, forced migration, disasters produced by anthropic action, etc., we must first and foremost consider the perspective of the victims, who are impoverished people.

The experiences and struggles of impoverished people must be given the first and principal voice, recognizing not only their marginalization, their invisibility (and invisibilization), but overall, their human dignity and their right to take place in the debate, participate in the analysis, collaborate in the comprehension of problems, and especially their right to decide on the issues they are most affected by.

Listening to and integrating victims' voices, perspectives, and rights (generally impoverished people) is a considerable challenge (not theoretical, but existential). It is not easy, because it implies escaping the ideology we are accustomed to inhabiting while also becoming existentially open to the voice, the truth, the perspective, the opinion, and the position of "others". This means promoting an inclusive and constructive dialogue that seeks sustainable and respectful solutions for those in trouble (in poverty in all its variations), ensuring that all human beings can live in dignity and participate fully in society, regardless of their economic situation.

I am not advocating against refining scientific instruments, methods, and results. Still, all these necessary developments acquire and express their best meaning when they are at the service of the dignity of every human being. If this is not the case, as a Jesuit of the last century asked, "*what is the value of putting a foot on the moon when it does not seem possible for all of us to put our feet on the ground?*" (De Mello 1990; emphasis added).

13.2 An Ethical Standard and Principal of Public Action

Ordinarily, we see and interpret the world from our places of comfort: where we eat and sleep. Yet now, we are gathered here at this conference for a few days, talking about scientific research, transdisciplinarity, poverty, inequality, injustice, and other related challenges—sleeping and eating in a country with one of the highest human development indices in the world. Mentioning this reality might seem impolite if we were not invited here to get to the bottom of things and called to imagine and build open paths to make our work effective. We seek to produce not only knowledge but hope, to benefit not only the academy but to translate knowledge into genuinely better living conditions for the poor.

Why should the people in the favelas of Rio de Janeiro or Manaus, the women working in the rice paddies of Bangladesh, or the tobacco fields in southern Africa *interest* me? Why should I be *interested* in the millions of people forced to migrate (by war, climate change, political situations, violence, etc.), or the Indigenous people from the Amazon, or the homeless people sleeping on the streets of our western cities? Do these people *interest* me? That is to say, are they part of my "*inter*" (internal) "*esse*" (being)? Are they in my field of concern? Does their situation hurt me? Does it outrage me? Are they a constituent part of my research dynamics and my life?

While I was a philosophy student in Bogotá, I discovered in the walls of the Jesuit house a sentence that has accompanied me for years and is a source of inspiration, including in this moment: "The world is not good if it is not for everyone." More than a philosophical principle, I propose it as a criterion for living and for acting: "The world is not good if it is not (good/suitable) for everyone." This is a communitarian ethical standard, a principle of public action, not of individual or private nature. It is not a conceptual or essentialist *telos* (philosophical, moralistic, or theoretical), but a concrete, practical action track. The sentence quoted above is an existential invitation to consider that what we do—beyond the moral imperative of goodwill—is really good when it produces life, liberty, rights, joy, health, and hope for those who have none of these things. And that if what we do produces or reproduces (consciously or unconsciously) death, exclusion, oppression, indignity, slavery, impoverishment, dehumanization, individualism, marginalization, inequality, misery, etc., then what we do (even out of goodwill) is *not* good, and so it is not good to continue doing it.

Therefore there is always a serious question about what we are doing and how we are doing it. Doing science and asking questions while sitting in a wealthy country is already a political posture. While it may be that we are genuinely and deeply interested in the poor, our work may end up reversing its trajectory—as a boomerang does—and reproducing the conditions that have excluded, impoverished, and marginalized them.

It seems necessary to me to question radically how current sciences proceed. If we do not change our usual methodological, epistemological, and ethical practices, there are severe risks that they will continue to remain at the service of those who have led the world to the cliff we are now on paving the way to a hyper-concentration of power in a mere one per cent of the world population.

13.3 Requirements and Challenges for Good Transdisciplinarity

How can we engage in academic discussions, discourses, texts (including this), and scientific products in such a way that they support something good?

(a) We need to recognize the existence, validity, and necessity of other knowledge and dare to unlearn, relearn, think, and rethink our ways of proceeding.

The ecology of knowledge cannot be culturally endogenous. If it wants to be authentically ecological, it must be open, interestingly curious, and riskily provocative. As Boaventura de Sousa Santos has said: "There is no global social justice without global cognitive justice" (de Sousa Santos 2014). Cognitive justice calls for equity in recognizing and distributing knowledge, challenging hierarchies of expertise and advocating for the inclusion of diverse ways of knowing.

(b) We need to recognize and be aware that historically, science has been influenced by specific socio-economic and political factors that have also shaped its further developments. Funding and corporate interests influence research priorities and outcomes. The possibility of building scientific communities independent of large capital and multinational corporations is a major issue. There is a growing recognition of the need for a more inclusive and diverse science that can address global challenges without being constrained by economic and, therefore, political partisan interests, making humanity's most pressing problems the focus of current research.

To counterbalance this inadequacy, several models have been proposed and explored to foster scientific independence. For example, open and collaborative science, or citizen science, which promotes free and shared access to scientific knowledge as a way of democratizing knowledge and its reproduction, reduces dependence on traditional funding. In doing so, it promotes a certain democratization of access to scientific knowledge and fosters collaboration between professional and amateur scientists all the while involving people and communities traditionally considered "non-scientists" (but perhaps with other knowledge) in research, thereby broadening the base of participants and perspectives.

(c) Participation in international collaborative networks can contribute to a fairer balance in the production of scientific knowledge, when combined with both support for scientists in developing countries and the open perspective of an ecosystem of knowledge and expertise beyond classical Western patterns. Such networks allow the sharing of resources, knowledge, and skills, while also helping us overcome the barriers imposed by funding and corporate control.

As a global public good, science must be accessible to and benefit society while recognizing the role of both private and governmental sectors.

In summary, although there are significant challenges, the possibility of breaking with the Western paradigm and building more autonomous scientific communities is a challenging but achievable goal. It requires a collective effort and an ethical commitment to research, innovation, transparency, and collaboration.

Over time, these efforts could lead to a more equitable scientific system that is representative of a diversity of global interests and needs.

To conclude, there are several challenges on the way to creating more just scientific communities:

- Connecting academics, practitioners, decision-makers, and concerned communities sharing knowledge and resources.
- Involving affected communities (victims) in the research process, implementation, monitoring, and control.
- Developing communication strategies that transcend disciplinary jargon and promote mutual understanding of different disciplines and levels of intervention.
- Fostering a culture of adaptability and resilience within research and practitioner teams to face political, social, and economic challenges.
- Building and finding flexible and sustainable funding, thereby allowing for the exploration of new ideas and adaptation to changes during the research process.
- Developing evaluation systems that consider interdisciplinary projects' complexity and long-term impact beyond traditional indicators.
- Building social networks for learning, collaboration, implementation, and monitoring.

References

De Mello, Anthony (1990) Awareness: the perils and opportunities of reality. Image Books. New York

Sousa Santos, Boaventura de (2014). Epistemologies of the south: justice against epistemicide. Routledge. London and New York

Roberto Jaramillo Bernal, S.J. Ph.D. in Social Anthropology (EHESS), is the current Secretary for Social Justice and Ecology at the general curia of the Society of Jesus in Rome. His contribution is rooted in the experience of accompanying indigenous peoples and poor communities in his pastoral work, especially in Colombia, Brazil, and Perú.

Chapter 14
The Challenges of Development and Well-Being: A Testimonial from Guatemala

Mayra Consuelo Leiva Tezó

Abstract This contribution examines the empowerment of Indigenous Maya K'iche' women and communities in Guatemala, through the personal and professional journey of Mayra Consuelo Leiva Tezó. It addresses the question of how community-based economic, social, and political strategies rooted in Indigenous identity and a solidarity economy can contribute to sustainable development and gender equity. This text served as the basis for a presentation at the International Biennial Conference on Environmental Justice 2024. The first part presents the personal background and formative experiences that led to Mayra's engagement with Indigenous identity, women's rights, and the struggle against exclusion and discrimination. The second part focuses on her work in supporting and strengthening community organizations for economic empowerment through education, entrepreneurship, and solidarity economy initiatives. The third part highlights community-led approaches to territorial management, emphasizing environmental sustainability, participatory planning, and innovative monitoring tools. The fourth part discusses the institutional commitment to building alternative economic models that resist colonial and neoliberal systems, aiming at social justice, land access, and democratic participation. Finally, the fifth part emphasizes the importance of gender equity and intercultural participation, envisioning a just society in which women's rights and contributions are fully recognized across all spheres of life.

Keywords Women's empowerment · Indigenous peoples · Sovereignty · Social and solidarity economy · Social markets · Alternative marketing · Sustainability

M. C. L. Tezó (✉)
Mayra Consuelo Leiva Tezó, ASERJUS, Quetzaltenango, Guatemala
e-mail: mercadoslocales@serjus.org.gt

T. Karcher SJ et al. (eds.), *Making Hope Possible*, Environmental Humanities: Transformation, Governance, Ethics, Law 7,
https://doi.org/10.1007/978-3-032-17719-3_14

14.1 Introduction

I am Mayra Consuelo Leiva Tezó, currently 48 years old, married to my husband Felix Salvador Santos Racancoj, both of us of Maya K'iche origin from the municipality of Quetzaltenango. We have a 16-year-old daughter. I come from a Maya K'iche family, consisting of my parents, Miguel Angel Leiva Boj (RIP) and María Consuelo Tezó, who formed a family of six children: Henry Otoniel, myself Mayra Consuelo, Vilma Patricia, Marleni Elizabeth, Miguel Giovani, and Fabiola Marisol, all with the surname Leiva Tezó.

14.2 The Chance to Represent

As a woman, I can say that I had the great opportunity to achieve professional success. I completed my secondary studies, graduating as a Legal Accountant from the School of Commercial Sciences in Quetzaltenango. Later, I began my studies at the Western University Center of the University of San Carlos of Guatemala, where I completed my undergraduate degree in Business Administration. One of my most important life experiences was in 1996, at the age of 19, when I participated in a youth Indigenous initiative called *U'mial Tinimit re Shelajuj N'oj* (Daughter of the People Under the Ten Wisdoms). I had the good fortune of winning and representing the *Maya K'iche*' youth of Quetzaltenango for a year (1996/1997), a space that allowed me to learn more about the situation of not only the Indigenous youth but especially Indigenous women and the native peoples of our country, Guatemala, facing exclusion, discrimination, and racism.

This experience brought me closer to recognizing and valuing my origins and learning much about the culture I belong to, appreciating elements such as clothing, language, knowledge, and the worldview of our people. I had the opportunity to attend various cultural, social, and sporting events and to get to know our national territory. I also attended meetings, workshops, and conferences related to Mayan cosmogony and the political, economic, and participatory situation of the Maya people in our land. From there, my knowledge and interest grew, leading me to participate in spaces that allowed me to promote and contribute to the participation of women and the Maya people in general.

14.3 Empowerment of Community Organizations

In 2009, I was given the opportunity to join the workforce of the SERJUS institution, where I have been developing my capacities and knowledge toward the economic empowerment of Indigenous *Maya K'iche* women through processes of supporting community organizations in at least 15 municipalities in western

Guatemala (Cabricán, Huitán, San Martín Sacatepéquez, San Carlos Sija, Concepción Chiquirichapa, San Antonio Ilotenango, San Lucas Tolimán, among others). These community organizations work with about 20 community groups consisting of approximately 400 people, with an average of 300 women and 100 men.

Since 2023, I have been enrolled in the Master's Program in Social and Solidarity Economy Business Creation and Development (MaCDESS) at the Iberoamerican University (IBERO). This program is conducted within the International Center for Research on Social and Solidarity Economy (CIIESS), where I am part of the third generation. This is made possible by the combined work of ASERJUS, IBERO, CIIESS, and the Comparte Network learning community, from which we have received methodological and capacity-building contributions that strengthen the community work we do.

My current work focuses on the development of the community entrepreneurship and commercialization component with organizations from the municipalities of Santa Cruz del Quiché, Chichicastenango, and Cantel. The work focuses on training and educating men and women from these municipalities' organizations on: food sovereignty, community economy, citizen participation, the economic, social and cultural rights of women, the development of solidarity collective enterprises, the implementation of collective commercialization spaces, and the development of community savings methodologies, among other activities.

All of the above is aimed at strengthening the institutional commitment to the reconstruction of community organization and its networks at all levels, from municipal to national, with the full participation of both women and men. This facilitates the economic, social, cultural, spiritual, and political reconstitution of our people, by the promotion of social relations based on justice, gender and generational equity, as well as balance and harmony, free from dispossession, colonialism, and its effects.

In my professional role, my responsibilities lie in strengthening the capacities, structure, and operation of community organizations, and in promoting alternative economic models based on solidarity and collaborative work. This approach not only enhances the economic self-sufficiency of the communities involved, but also encourages their active participation in managing themselves.

At SERJUS, we have developed comprehensive training and education programs that instruct men and women of the organizations in key areas such as organizational management, entrepreneurial skill development, and principles of the solidarity economy. Furthermore, we work closely with these organizations to establish functional internal structures, drafting bylaws and participatory decision-making systems that ensure full engagement. In operational terms, we provide simple tools and methodologies to optimize the daily activities of organizations, from strategic planning to responsible financial management and social impact evaluation. At the same time, we encourage the adoption of innovative economic models, such as connecting and linking with other organizations and networks, which not only strengthen community cohesion but also drive local economic development.

14.4 Territory Management in the Focus

A crucial aspect of my work is empowering community organizations, particularly women, to meaningfully participate in territory management. This includes advocating for equity and environmental sustainability approaches in territorial planning, protecting local natural resources, and promoting practices that ensure long-term well-being, contributing to the care of our common home.

To ensure the effectiveness of our interventions, we aim to establish monitoring and evaluation systems that measure the impact of our actions using tools and methodologies such as the Participatory Guarantee System (SPG), Solidarity Economic Circuits (CES), and Socio-Environmental Audit System (SASA), among others. These systems allow us to adjust strategies as necessary, ensuring that our initiatives are not only successful in the short term but also contribute to sustainable and lasting community development.

In our work with community organizations, in dealing with training and capacity-building, we follow the following steps:

1. Political and Technical Training in food sovereignty and community economy.
2. Development of Collective Enterprises from the solidarity economy.
   - 2.1 Sustainable extraction: for collective benefit.
   - 2.2 Production: agroecological, food-based.
   - 2.3 Transformation: artisanal, semi-industrial, associative.
   - 2.4 Marketing: local, exchange, barter, direct producer–consumer relationship.
   - 2.5 Consumption: responsible and solidary.
3. Promotion of Local Farmers' Markets: consumer awareness campaigns.
4. Economic Empowerment of Women: community savings and credit methodology.
   - 4.1 Community savings for women.
   - 4.2 Women's access to credit.
5. Training for Public and Political Management and Advocacy at the local, regional, and national levels.
   - 5.1 Development of economically productive projects from community organizations for management by municipal authorities.
   - 5.2 Support for national organizations in their work to gain the approval of legislative initiatives that care for our common home.

14.5 Institutional Commitment

Operationally, everything described above is carried out through the facilitation of tools and methodologies intended to optimize the daily functioning of organizations, from strategic planning to responsible financial management and social impact evaluation. At the same time, we adopt innovative economic models, connecting and linking with other organizations and networks, which not only strengthen community cohesion but also drive local economic development. In this, we remain focused on our institutional commitment to:

- The social and political struggle against colonialism and ethnic and class oppression and for the democratization of access to land, water, and income.
- The construction of a pluralistic, just, and democratic legal-political system.
- Overcoming the disconnection of social forces, especially the dispersion and fragmentation of the vast exploited, dominated, and impoverished majorities, while seeking to elevate their capacity for proposal, planning, participation, and action so they can weave their own development.

By doing so, we aim to create communities with capacities and favorable conditions to face the effects of the economic crisis, the increase in poverty, and the food crisis. This promotes the strengthening of local community economies and the involvement of municipal governments with citizen participation in fostering economic growth. These efforts are geared toward local needs and demand, based on mutual aid, fair trade, microfinance, sustainable agriculture, and agroecology, as well as respect for Mother Nature, humanity, and life, as an alternative to the profit-driven sensibilities of dominant economics.

14.6 Gender Equity and Participation

On a personal level, my commitment is to contribute to building a society with justice and gender equity, primarily ensuring real and effective participation by the four ethnic groups who cohabit this territory of Ixim Ulew (Land of Maize), Guatemala: the Maya, Garífuna, Xinca, and Mestizo peoples.

With the full participation of these peoples, I aim to ensure that the work of women is recognized and valued, that household tasks become a team effort, and that machismo, the attitude of considering men superior to women, has no place in the lives of either men or women. I dream of a society where women can participate freely and exercise their rights without overburdening themselves with work and without fear of being undervalued.

I hope to see women's economic contributions recognized, increasing their economic, political, and social participation from the community level to the municipal, regional, national, and international levels.

In summary, my work focuses on training, strengthening, and actively supporting community organizations in building an economically and environmentally sustainable future, where solidarity and citizen participation are fundamental pillars of development and well-being: 'el buen vivir' in Spanish, in Maya K'iche', *Utz K'aslemal.*

Mayra Consuelo Leiva Tezó A Maya K'iche' community leader and business administrator from Quetzaltenango, Guatemala. She currently pursues a Master's degree in Social and Solidarity Economy Business Creation and Development at the Ibero-American University. Her work focuses on the economic empowerment of Indigenous women, food sovereignty, community entrepreneurship, and participatory local development in Guatemala.

Chapter 15
Indigenous Knowledge and Environmental Guardians: A Testimonial from the Philippines

Jason Linantad Menaling

Abstract This contribution explores how Indigenous knowledge systems contribute to environmental guardianship, based on the testimonial of Jason Menaling from the Pulangiyēn community in Mindanao, Philippines. It addresses the question of how traditional ecological knowledge, land rights, and cultural continuity support environmental justice and sustainable development. Originally developed as the basis for a presentation at the International Biennial Conference on Environmental Justice 2024, this chapter presents the Pulangiyēn worldview and ancestral knowledge as foundations for environmental guardianship and intergenerational learning. Particular attention is given to land rights and the significance of the Ancestral Domain Title as a legal and cultural instrument for sustainable land management. The chapter further addresses the divisive effects of national economic development plans and market pressures on Indigenous communities and their livelihoods, as well as the impact of climate change on traditional agriculture, food security, and ecosystem balance. Collaborative approaches to socio-ecological transformation are ultimately emphasized, highlighting education, community empowerment, and culturally informed resource management.

Keywords Indigenous communities · Environmental guardianship · Intergenerational learning · Landrights · Traditional agriculture · Food security

15.1 Introduction

I am Jason Menaling from the island of Mindanao, Philippines. I grew up by the Pulangi River in Bendum where my parents provided for eight of us children from diverse food crops on our small farm lots by the forest and through some basket and mat weaving. I continue to live with my family in these ancestral lands planting our

J. Linantad Menaling (✉)
Malaybalay City, Philippines
e-mail: jasonmenaling@essc.org.ph

T. Karcher SJ et al. (eds.), *Making Hope Possible*, Environmental Humanities: Transformation, Governance, Ethics, Law 7,
https://doi.org/10.1007/978-3-032-17719-3_15

own food crops while working as the field program manager at the Apu Palamguwan Cultural Education Center (APC). I appreciate this chance to speak today. As an Indigenous person from a marginalized group, my involvement here is part of my effort to share about our way of life, culture, and our relationship with nature.

For us Pulangiyēn, our ancestors have long valued and protected our environment received from the Creator. They passed down knowledge about managing natural resources, which helps us today. This knowledge, shared by our elders, teaches us how to care for our environment and pass these lessons to younger generations. Institutions like APC that run a kindergarten-to-grade-12 school in the mother tongue play a key role in educating our youth about our cultural heritage and environmental guardianship. This knowledge is a legacy we aim to sustain for future generations.

15.2 Land Rights and Environmental Justice

For generations, our ancestors lived on and cultivated their land with care, never overexploiting it, and giving time for the land to rest and regenerate. It's crucial that their efforts to protect and claim their land are respected, especially by our government. Economic interests should not overshadow the rights and traditions of Indigenous peoples.

The Ancestral Domain Title was granted in 2014 after more than 20 years of logging and a further 26 years of negotiation. It is the smallest Certificate of Ancestral Domain Title (CADT), comprised of less than 2000 hectares of 60% premontane and montane forest. This is more than a piece of paper for a legal claim: it represents our right to protect and manage our land according to our cultural practices. Our elders' stories and knowledge are essential for understanding and maintaining this land. We have traditionally grown crops like hill rice, sweet potatoes, cassava, local corn, millet, and yams. We have only cut trees for our houses and leave our fields in the forest to regenerate.

The lower elevation lands reaching the Pulangi River were logged over forty years ago and burned by the migrants from nearby islands, who came with the company and who were also poor and needing land. Today, we have been very fortunate to protect our water source, yet share it with everyone and filter it back into the land. Our water is good for the city below and the agriculture as we do not use chemicals.

Government needs to pay for permanent local forest managers who insure the integrity of the forest and provide valuable ecological services. This is part of the justice we seek in society. The air, water, and biodiversity should be enough for the government to recognize the value of community social fencing and forest management.

15.3 Divisive Impacts of National Economic Development Plans

Sadly, one of the more destructive agricultural practices has been the push to grow high-yielding varieties (HYV) of maize for commercial animal feeds, driven by aggressive market distributors, traders, and politicians. Our roads are not built as "farm-to-market" roads for small farmers to bring their produce to the local markets, but for traders with preset cropping arrangements to gain access with trucks of 16–25-ton capacity to take the harvest from communities who are not familiar with the ways of the market.

Many Indigenous communities have lost their land to large operations due to financial and political power. This threatens our way of life and our ability to manage our land sustainably. We are the last village in the valley that manages its own sustainable water system through a Tribal Council, rather than through local government officials. We need our government to recognize and support our traditional practices, and to ensure that our environment is managed for future generations.

It is hard for the culture and for families to balance their actual needs with what they see others have and their wish to have it. Of course, the youth want the technology and the entertainment and to ride around on motorbikes. When there is so much attraction and promise in the city, these immediate wants are hard to balance against crop diversity and maintaining productivity on the land. So, a job as a security guard or packer might be fine for a time as a single person. A salary gives the person what they immediately want, but minimal money and inflexible time schedules do not work for a young family—or the need, for example, for hospital visits. The wants of the family often go beyond their needs, and this pushes them to seek loans that many times they cannot manage. Payment of the interest may force the family to sell any land they or the broader family have.

Government needs many more programs to empower local communities to develop the skills for a sustainable engagement with the market and risk management.

15.4 Climate Change Impacts on Indigenous Communities

People in remote areas, particularly Indigenous communities in the Philippines, are feeling the effects of climate change. Changing weather patterns challenge us all and require collective solutions. In the past, our elders managed resources effectively to ensure food security, but unpredictable weather now complicates this. The planting season can vary and the starting point is not always clear. Furthermore, with the commercial corn, rats are now a major pest, making it unproductive to grow hill rice, and the herbicides have allowed for new species or grass to invade.

The loss of crop diversity, along with changes in temperatures, rainfall, moisture, and ecosystem conditions, have all made it harder for us to maintain our traditional ways of life. We face increased difficulties with food security and adapting our farming practices. The uncertainty of the climate affects our trust in nature and impacts our daily lives. We must address these challenges together and find ways to adapt while maintaining our cultural practices.

15.5 Collaboration and Empowerment for Socio-Ecological Transformation

In the region of Pantaron Range in Mindanao, local efforts are focused on sustainable management of natural resources through Forest, Farm, and Leadership in the Margins (FFLM), a farm school working on organic food production, bamboo construction, and citronella oils that also contributes technical courses in the high school. Empowerment comes through the language and collective understanding and memory of the culture. For example, this includes understanding the various interrelated species in ecosystem regeneration and ensuring long-term access to clean water. We have documented 300 plant species that the children learn as a progression in forest regeneration and water filtration. APC, operated by local teachers for 400 children in elementary school and high school, works together with the Institute of Environmental Science for Social Change (ESSC) to engage our youth in cultural and environmental management.

The Ancestral Domain Sustainable Development and Protection Plan (ADSDPP) guides our natural resource management, forest, biodiversity, food, water, and, to a lesser extent, cash crops, with cultural principles led by the Tribal Council and the people of Gaup in Bendum. We have an educational program that responds to our cultural integrity while giving new opportunities for tertiary education through Jesuit Worldwide Learning's professional diplomas and academic degrees. We discuss the need for meeting our own food needs and share about the risks. While looking for the best opportunities for growing cash crops, we work collaboratively with neighboring communities to ensure that our efforts benefit everyone. It is essential to support each other and to protect our environment for the well-being of all, and this deepens our sense of shared hope for the land and coming generations.

Jason Linantad Menaling Datu of the Bendum Pulangiyēn Ancestral Domain, head of the tribal council and program manager of Forest, Farm and Leadership in the Margins, Apu Palamguwan Cultural Education Center and student of Jesuit Worldwide Learning (JWL).

Chapter 16
Sustainability in the Making: The Bernese Consensus on Education for Sustainable Development

Robert Unteregger

Abstract This chapter presents the Bernese Consensus: a project that formulates a minimal consensus on how Education for Sustainable Development (ESD) can be taught in schools today. The consensus is inspired by the Beutelsbach Consensus on political education, formulated in the 1970s. There are similarities in method and goal, but differences in the object (political education, ESD) and in the ethical position. The original minimal consensus, focused on enabling pupils to analyze and act according to their own political interests, must evolve to incorporate the broader range of societal challenges and responsibilities outlined by sustainable development frameworks such as the UN Sustainable Development Goals (SDGs). In accordance, the ethical position must be a different one, too. The chapter requests that learners must be taught in a way that motivates and empowers them to get involved actively in shaping a sustainable future. This is considered a core element of ESD. It shows how ESD is by now anchored in education policy and curricula, especially in Switzerland. It also shows the difficulties of ESD on its way from concepts to becoming practice. There is no straight way to take theory to schools and classrooms. The chapter describes how the initiative of the Bernese Consensus on ESD as a practical response to this obstacle has evolved, proposing five key pedagogical principles on how to teach ESD in schools: connectedness of head, hands and heart; exemplary teaching of themes; pluralistic, activating methods (interdisciplinarity, projects, personal development, cooperation, participation …); whole school approach; and fostering attitudes of modesty, perseverance, imagination, and hope. Finally, the chapter argues that ESD, facing and tackling the challenges of sustainability with open eyes, grounded in democratic values and implemented effectively in schools, can empower and motivate young (and elderly) people to address those global challenges through coordinated, persistent action, thus making hope possible.

R. Unteregger (✉)
Pedagogical University of Berne, Berne, Switzerland
e-mail: stiftung@zukunftsrat.ch

Future Council Foundation, Berne, Switzerland

T. Karcher SJ et al. (eds.), *Making Hope Possible*, Environmental Humanities: Transformation, Governance, Ethics, Law 7,
https://doi.org/10.1007/978-3-032-17719-3_16

Keywords Education for sustainable development ESD · Bernese Consensus on ESD · Beutelsbacher Consensus on political education · Didactical principals for ESD · Minimal consensus · Common sense · Whole school approach

16.1 A Historical Inspiration

Within the context of the Cold War and the political competition between Left and Right in Germany, and particularly in the state of Baden-Württemberg in the 1970s, there was a pressing societal and educational issue. That is, society sought to address how political education (PE) ought to be handled and taught. Was PE's aim to learn to change the existing political system, as the rather left-oriented teachers would propose? Or would it consist in strengthening the existing order of the state, as the rather conservative teachers intended? At a conference in Beutelsbach, those in charge of education met to discuss this very issue. Following the conference, a minimal consensus on how to teach PE was formulated by Hans-Georg Wehling for the documentation of the conference, consisting of three main ideas:

1. It is forbidden to overwhelm pupils with one's own opinion or position and to hinder them from forming their own opinion. This is exactly where the border between PE and indoctrination is situated. Indoctrination is incompatible with a democratic society and with the aim of the people's majority.
2. What is controversial in society, science and politics, must be presented as controversial and from different perspectives.
3. The pupils are to be enabled to analyze a political situation and their own interests in it, and they should be given space to explore various ways and means for how they can influence the political situation according to their best interests.

These few and succinct key ideas regarding how to teach PE have provided a very helpful orientation for many teachers and people in charge of education since the 1970s until today. The leading idea was to create a *minimal* consensus with as few succinct key points as possible. And these points were focused on the perspective of how teachers should teach PE to their pupils in the classroom.

16.2 Differences in Range (SD) and Ethical Position

When two lecturers on ESD from the Pedagogical University of Berne and from the Pedagogical NMS University of Berne started to formulate the Bernese Consensus on Education for Sustainable Development (ESD), the idea of creating a minimal consensus with as few and succinct key sentences as possible was very helpful and instructive. On the other hand, we realized that the aim of PE as it was formulated in the 1970 is incompatible with the aim of ESD in our time, in the 2020s. The final aim of PE was, at the time, to enable the individual to analyze their own interests in a given political situation and to influence it according to those interests. Today, in the context of ESD, it doesn't do to reduce the perspective to one's personal positions, actions, and interests. We have to take into account the perspective of sustainable development, too, for example, by applying concepts like the United Nations' Sustainable Development Goals (SDGs), the Doughnut Model of Kate Raworth, the planetary boundaries, or others. Being aware of sustainable development perspectives and of one's own interests, one has to create and develop new ideas, opinions, and positions, finding a viable balance between the sustainable development goals and one's own interests.

Considered from the perspective of a human, civil, and democratic society, our society, the world, and especially the future generations are threatened by several overwhelming powers:

- The effects of the Anthropocene: The term was introduced by the scientists Paul J. Crutzen and Eugene F. Stoermer in the year 2000 to name a new geological age where the influence of humanity on the development of the earth is stronger than natural changes. One of its effects is the climate crisis.
- The system of the worldwide market society, where the economic sector, disembedded from society, tends to destroy society and nature because of its unsustainable concepts of development (unlimited economic growth, waste of natural resources).
- Numerous wars and conflicts.
- Rising populism and authoritarianism that tries to legitimate itself with formally democratic instruments.
- New communication techniques and digitalization, intruding into one's personal sphere (risk of flooding, bubble-building, and fake news).
- Interculturality: Through globalization and worldwide mobility, the population consists of as many cultural elements and languages as ever before. That offers many possibilities, but it is also a huge challenge for living together.
- Complexity and speed of many social, technical, and ecological developments.

That non-exclusive list may serve to illustrate the ways the lives of many citizens, and especially young people today, are influenced and dominated by a large number of dependencies and overwhelming forces.

16.3 (Re-) Gaining Steering Power, from a Pedagogical Perspective

On the ground of our common, inalienable rights and freedoms,, and on the value basis of our human, civil, democratic society (which, of course, isn't perfect at all, but which nevertheless is viable in a human and practical way), we are to live this community of values, to shape, and to develop it further, under the guiding umbrella of the constitution and the UN-Agenda 2030. ESD aims at strengthening the pupils' ability to reach majority, to shape one's life as a responsible citizen, and to assume responsibility in our society and for its development. To teach those goals, and in response to the above-mentioned overwhelming forces, ESD must strengthen and encourage learners' resilience, orientation, capability to act and self-reliance, in the following ways:

- Giving and Fostering Orientation: Exemplary, by contact, by getting to know, knowledge, by forming one's opinion and positions, developing attitudes, exploring and exercising (new) ways of acting.
- Knowing and Acting: From the beginning, knowing and acting must be in contact and balance. Much knowledge without action leads to frustrated knowledge, and acting without knowledge ends in blind activism.
- Modesty, Perseverance, Imagination, Preserving Hope: As individuals, and in view of the dimensions of the challenges of sustainable development, we are no more than small drops in an ocean, and yet we are clever, thoughtful, acting, and networking drops. Acting in coordinated ways, together with millions of other small drops, with imagination and will, we can—as foaming waves and as calm sea—create big things and sustainable development.

Such a value-, knowledge-, and action-based approach to teaching differs from today's widespread teaching, based on a list of competencies and following the orientation of the policy proposed by OECD's DeSeCo-papers (*De*finition and *Se*lection of *Co*mpetences; Rychen and Salganik 2003, OECD 2005). There the aim of education is flexibility and adaptability to the manifold challenges of modern society. It differs also from a teaching where the solutions to the challenges of SD are considered as known and the resulting teaching is understood as a training in knowledge and skills to realize the known solutions. ESD is a process where the learners can be, to a certain degree, co-creators and co-shapers of a sustainable future. Teaching should allow learners to explore different ways of action toward SD, to form and develop their own abilities, attitudes and positions, and to act and take on responsibility. That learners can do that actively is a core element of ESD. Futures are not understood as a given or as something totally out of reach, but as something to a certain degree open and to be shaped actively (Holfelder 2019).

16.4 ESD in International and National Education Policies and in Swiss School Curricula

Today, Sustainable Development (SD) is, at least formally, accepted as a central objective by a political majority of the population. It is part of the main set of documents that guide societal, political, interpersonal, and international life and cooperation. SD is formulated in the UN-Agenda 2030, in the Swiss Constitution, and in the Strategy SD 2030 of the Swiss government. Education is considered a central lever for advancing SD.

SDG 4, section 7 of the UN-Agenda 2030 contains a general definition of ESD: By 2030, ensure that all learners acquire the knowledge and skills needed to promote sustainable development, … (UN 2015).

In the Swiss Constitution (since 1999), aspects of SD are mentioned in several articles: promotion of common welfare, sustainable development, … committed to the long-term preservation of natural resources and to a just and peaceful international order (Article 2) … endeavor to achieve a balanced and sustainable relationship between nature and its capacity to renew itself and the demands placed on it by the population (Article 73) (Swiss Federal Authorities 1999).

In the Strategy SD 2030 of the Swiss government (since 2022), based on the above-mentioned documents and others, the federal government has formulated its Strategy SD 2030. It mentions education as one of four main levers for SD, together with civil society, economy, and financial market (Swiss Federal Authorities 2022).

ESD is mentioned in all the Swiss curricula for primary and secondary schools.

In the curriculum of the obligatory school of the Swiss-German speaking part of Switzerland (Lehrplan 21, since 2015), ESD is introduced with these words:

The pupils study the complexity of the world and its ecological, economic, and societal developments. They realize and understand the interconnectedness and the contexts and are enabled to participate in sustainable future shaping (Swiss Conference of cantonal directors in education 2015). In a separate chapter, ESD is introduced and defined more specifically, followed by some didactical key concepts on how it should be taught in school.

In the curriculum of Latin Switzerland (Plan d'études romand PER, since 2010), there are, in addition to the corpus of the traditional subjects, two new corpuses with their own wide themes: education générale including health and well-being, personal projects, living together and training in democracy, interconnectedness (social, economical, and environmental), and transversal competencies, including collaboration, communication, learning to learn (strategies), creative thinking, and reflection.

(Intercantonal Conference in public education, Latin Switzerland 2010)

In the Swiss curriculum of the grammar schools (since August 2024), ESD is introduced in a separate subchapter. It explains in an open manner how ESD is to be understood, focuses on the several transversal competences to be taught (corresponding much to the ESD-competences of UNESCO), and mentions that ESD must be taught in the single subjects, and by school culture and in a whole school approach, as well (Swiss Conference of cantonal directors in education 2024).

16.5 No Direct Way from Conceptual Framework to Practice

As the above-mentioned documents show, a conceptual framework for ESD in teaching and school-organization does exist. But there are many challenges on the way from didactical concepts to effective practice.

When we look at the UN-SDG, the range of possible themes that should or could be dealt with is endless. Where should we start, or how should we choose the few themes we could probably find time to work on in the very limited time windows at disposition?

In the curricula, every traditional subject has an assured number of lessons and a curriculum of its own. The objectives to be reached within an ESD are circumscribed most often by transversal themes without an assured number of lessons, and by very general competencies. The task of introducing them and combining them with the existing subjects seems a difficult task, since the school is still strongly based on the subjects and their lessons, and less on transversal themes, collaboration, and competencies.

Since the SDG themselves are a canon of values, it is obvious that ESD has something to do with building up and strengthening the ability to form opinions, positions, decisions, and the ability to act in a societal context—and all this with an orientation with respect to the SDG. ESD has something to do with forming attitudes, too, and with building up the resilience to tackle and work on challenges that we can solve neither on our own, nor in a short time.

In Switzerland, education is organized in a very federal way, given its several national languages, its 26 cantons, and its more than 2000 municipalities. For example, there is a common Swiss frame curriculum for the grammar schools. Once that curriculum is in force, every canton (or at least the larger ones) formulates on that basis its own curriculum for grammar schools. One of the cantons, St. Gallen, has developed its curriculum for grammar schools at the same time as the Federal authorities have developed the federal one, and paying attention to it, so that it would conform to the federal version. When the Swiss curriculum entered into force in August 2024, St. Gallen already had its corresponding cantonal version, while the other cantons will formulate theirs in the coming years. Most Swiss-German cantons will put theirs into force come 28/29, but the grammar schools of Latin Switzerland will do that only in the first years after 2030. The reason for this is that until now their grammar schools worked with only three school years, and now they must add an additional school year so that all grammar schools in Switzerland will last at least four years in the future.

Given this very federal structure of education in Switzerland, it is very difficult to get an accurate overview on the state of ESD in schools (and on other issues, too). Making use of the classification of Sterling to estimate to what degree a school is practicing ESD, from "denial," "bolt on," "built in" to "whole system redesign" (Sterling 2004) as an orientation, I suppose that one could find schools of all those degrees. The federal structure leaves to the single school and its headmaster and to local and regional institutions considerable space to maneuver.

16.6 The Initiative for a Bernese Consensus on ESD

In the department for grammar school teachers at the Pedagogical University of Berne, in the seminary on ESD, I can introduce students from all subjects to ESD. How to teach values and form attitudes, and the role of the teacher in such a process, has always been a central and lively discussion theme within the seminary on ESD. In 2021, when I prepared a unit on the key ideas of the Beutelsbacher Consensus, I realized that what was very helpful for political education in the 1970s and has been until today wouldn't do for teaching ESD today. The focus on enabling the individual to analyze political situations and their own interests in those situations and to give them space to explore ways and possibilities of how to advance their interests in those situations is incompatible with the structure of ESD. In ESD, we must analyze and take into account as well the perspective of SD, and we have to learn how to find a balance between our personal interests and the interests of SD, as already mentioned at the beginning of this text. We should be given space and time to explore ways and possibilities how to advance those balanced interests. Following this insight, I formulated correspondingly points 4 and 5, in addition to the three key ideas of the Consensus.

For a few years, the four Bernese universities (two pedagogical, one general, and a vocational one) have had a common forum on ESD called "ESD in Dialog" where their teachers engaged or interested in ESD meet and exchange and present experiences, ideas, and projects. In November 2024, Alexander Schroeter, a lecturer on ESD at the Pedagogical NMS University of Berne, made a presentation on the Beutelsbacher Consensus and its relations and non-relations to ESD today. At the end, he proposed that a new "Bernese Commitment" on ESD could be formulated. I was the only one who reacted actively to that proposition. Given my own reflections on the Consensus and ESD, and the huge challenge to implement ESD effectively in Swiss schools, I thought this was a very pertinent and urgent project. It could help to advance ESD at the Pedagogical Universities and to give orientation for an effective practice of ESD to the teachers in the schools—as well as publicly.

We met in January 2025, and by February 12 we had elaborated a first draft for a Bernese Consensus on ESD. We sent the draft to all the participants of the forum "ESD in Dialog" and to some colleagues of ours engaged in ESD. We asked them for feedback. By March 24 we received twelve answers, all but one critical-positive, a few even enthusiastic. We integrated the feedback until March 26. In June, we added a small annex, explaining how the text had come into being and how to make use of it. This is, as of July 2025, the text of the Bernese Consensus and the well-developed basis on which we plan the next steps of this project. On one hand, we will discuss the draft with further interested colleagues in Berne, Switzerland and abroad; on the other hand, we will contact educational institutions and political authorities, inform them and invite them to make use of it. As soon as possible, we will install easy access to the text on an internet platform, so that the process is transparent and understandable to all interested in it. We trust that the consensus will make its way by being discussed, used, and applied.

The introduction to the draft of the Bernese Consensus contains many of the above-mentioned notices on ESD. It ends with five pedagogical key points, focusing on how to teach and organize ESD in classrooms and schools:

1. Head, Hands, and Heart

The whole human being is required! In the short subchapter on ESD in the frame curriculum for grammar schools 2024, it is mentioned twice that ESD requires the fostering of all human faculties. It concerns "the teaching of knowledge, skills, values, and attitudes," and the "cognitive, social, emotional as well as the behavioral dimensions of learning" are to be strengthened.

2. Content: Exemplary

The UN-Agenda 2030 with its 17 SDG stands for a huge number of themes and fields of action. Not all can be taught, and those that are taught cannot be taken into consideration to the same degree. Depending on context, age, and character of a school, the main emphasis can be chosen differently. Some themes must be taught in exemplary ways and more deeply, the goal being to form a comprehensive understanding of what the SD of a society and of the world may look like, connected with an ability to act globally.

3. The Ways of Learning and Working in School

To make it possible for pupils to develop the competences required for ESD, the ways of learning and working in school are to be developed further and, to some degree, created anew: more project-oriented, interdisciplinary, transversal, self-guided, personalized, participative, and cooperative.

4. The Whole School is Required

ESD must not fizzle out by proceeding through only occasional events and actions. It must be organized and shaped comprehensively. Required are:

- an intelligent school, where the teachers collaborate and coordinate their work,
- the participation of the pupils, and,
- a comprehensive organization of ESD from the beginning to the end of the school-days and in a whole school approach. All the actors of a school participate and interact. In the frame curriculum for grammar schools, that approach is explicitly engrained.

5. Attitudes, Not Only Contents.

Self-reflecting modesty, perseverance, imagination, and keeping hope. Normally, pupils do schoolwork where they can find adequate solutions (solving an arithmetic problem, writing a composition, observing the behavior of an animal, learning vocabulary, …).

In the context of ESD, we learn to work on huge challenges that we cannot solve alone, and where our contribution to a solution can be only a very tiny part. But by working together—being foaming waves and glittering, calm sea—we can create SD. To learn to work continuously, persistently, with modesty, rich in ideas and with hope: that is a central task of ESD (translation RU).

Do ESD-schools structured along that minimal consensus already exist? Connecting the Future Building site at Lake Neuchâtel and the ESD seminary at the Pedagogical University of Berne, we were able to initiate the project "Pioneer grammar schools in ESD—implement ESD in a whole school approach" in 2019. On the 4th of July 2025, we could certify the first grammar school as a pioneer school in ESD in Switzerland, documented by a report of ninety pages (Unteregger 2025). There are other grammar schools participating, and a first lower secondary school is involved, too. We can realize that project in dialog with the Swiss Academies of arts and sciences and in collaboration with the members and secretariat of the Club of Rome. For the German-speaking countries, a handbook on ESD is in preparation and expected to have been edited in 2025 (Rieckmann et al. 2025).

On an international level, several recently provided comparable studies on the state of ESD in schools and teacher formation exist (i.e., Hooda and Tuba 2025, Mäki-Kutti et al. 2025, Wals and Mathie 2022a). Arjen Wals, professor of transformative learning for socio-ecological sustainability at Wageningen University and holding the UNESCO Chair of Social Learning and Sustainable Development, has been focusing for a long time on linking ESD with the concept of whole school approach (Wals and Mathie 2022a, b). A product of the Erasmus KA2 program (2016–2018) has been the booklet "A Roadmap to an ESD-school" (Saloranta 2018). The project was based on developing ESD in four European schools for several years in a whole school approach. Its orientation corresponds quite fully with that of the consensus.

On a global level there exists a strong tradition of eco-schools, based in environmental education. In the worldwide project "Eco-schools" of the Foundation for Environmental Education (FEE), started in 1994, there have been more than 50,000 schools and more than 13 million pupils involved (https://www.fee.global). Among many educational organizations originally focusing on environmental education, there exists a tendency to connect and combine with ESD.

Overall, we may say that there exist quite some efforts to promote the integration of ESD in schools, in Switzerland as well as abroad. I am in contact with some of the above-mentioned pioneers in ESD, and we are reflecting on how to connect and to coordinate our efforts to strengthen and advance a well-structured building of ESD in schools. During the last decades, a solid common ground has evolved. We are not looking for a golden formula, but rather to strengthen a growing, emerging way of practicing ESD: well-structured and effectively, in whole schools as well as in teacher education.

16.7 A Perspective—Making Hope Possible

"How can hope be made possible by action-oriented transdisciplinary collaboration and research?" is one of the guiding questions of this volume.

On our way out of the Anthropocene following the guidance of the SDGs and the Doughnut (or lifebelt) of a just and safe space for humanity that includes the regenerative and distributive economy of Kate Raworth, implementing ESD in our schools in a comprehensive and effective way is a crucial lever. This insight is formulated in the Strategy SD 2030 of the Swiss government, as well as in target 4.7 of the UN-SDG on education.

The way out of the Anthropocene in general requires numerous actions from billions of humans worldwide. When considered from the perspective of each individual human, the way out can only be found step by step and built brick by brick. But those steps and bricks taken together can produce immense and fast changes—as sparkling waves and a glittering, calm sea. Is this poetic? Yes, it is. Is this poetry? Yes, it is! Or, as the French say, this is "*une vérité à faire.*"

One important step or brick on this way is to bring ESD into our schools and to enable teachers and pupils and whole schools to participate in shaping a sustainable future while also forming our schools into laboratories for SD (UNESCO 2021).

To accept and shoulder part of this huge task in an intelligent path forward, and to keep working on it persistently makes hope possible, causing sparks of hope that may soon form a strong, lasting fire of hope.

Since a teacher in an obligatory school normally teaches several different subjects, there are often strong elements of transdisciplinarity and there may be elements of transdisciplinary in his/her teaching, too. Nevertheless, in grammar schools interdisciplinary and transdisciplinary teaching are especially rare. In these schools, teachers understand themselves traditionally as teachers of their subject (often one subject, sometimes two), corresponding to their formation as masters or doctors in their subject. In addition to their traditional role, they must now also reflect upon what their role in school should be given the challenges presented by the SDGs, so that they can tackle those challenges by collaborating with their colleagues and by using more interdisciplinary and transdisciplinary approaches.

The Bernese Consensus can help to raise awareness and advance effective implementation of ESD in schools using a whole school approach that furnishes a few understandable, general key points regarding how to teach and organize ESD. The Consensus can also thereby help to foster a new common sense, where SD and ESD are commonly understood and provide a strong and creative element of orientation.

References

Holfelder A-K (2019) Towards a sustainable future with education? In: The politics of making and un-making (sustainable) futures, vol 14. Springer Nature, pp 943–952

Hooda A, Tuba O (2025) Comparing sustainability competences across school curricula in Europe and beyond. Retrieved 24 July 2025 from https://inclusive-future.eu/wp-content/uploads/2025/06/SUSTAINABILITY-SCHOOL-CURRICULA_InclusiveFuture_May2025.pdf

Mäki-Kutti M, et al. (2025) A sustainability handbook and toolkit for schools, teachers and policymakers. University of Jyväskylä

OECD (2005) The definition and selection of key competencies: executive summary. Retrieved 24 July 2025 from https://www.deseco.ch/bfs/deseco/en/index/02.parsys.43469.downloadList.2296.DownloadFile.tmp/2005.dskcexecutivesummary.en.pdf

Plan d'études romand (2010) Inter-cantonal conference in Public Education, Latin Switzerland. Retrieved 24 July 2025 from https://www.ciip.ch/Instruction-publique/Plan-detudes-romand

Riekmann M, Singer-Brodowsky M, Bertschy F, Rauch F (to be published in 2026) Handbuch Bildung für nachhaltige Entwicklung. Barbara Budrich

Rychen DS, Salganik LH (2003) Key competencies for a successful life and a well-functioning society. Hogrefe & Huber

Saloranta S (2018) A roadmap to an ESD-school. Erasmus+ KA2 program (2016–2018)

Sterling S (2004) Higher education, sustainability, and the role of systemic learning. In: Corcoran P, Wals A (eds) Higher education and the challenge of sustainability: problematics, promise, and practice. Springer, pp 49–70

Swiss Conference of Cantonal Directors in Education (2015) Curriculum 21. Retrieved 24 July 2025 from https://www.lehrplan21.ch/

Swiss Conference of Cantonal Directors in Education (2024) Rahmenlehrplan gymnasiale Maturitätsschulen. Retrieved 24 July 2025 from https://www.edk.ch/de/themen/gymnasium

Swiss Federal Authorities (1999) Federal Constitution. Retrieved 24 July 2025 from https://www.fedlex.admin.ch/eli/cc/1999/404/en

Swiss Federal Authorities (2022) Strategy SD 2030. Retrieved 24 July 2025 from https://www.are.admin.ch/are/de/home/medien-und-publikationen/publikationen/nachhaltige-entwicklung/strategie-nachhaltige-entwicklung-2030.html

UNESCO (2021) Berlin declaration on education for sustainable development. Retrieved 24 July 2025 from https://unesdoc.unesco.org/ark:/48223/pf0000381229

United Nations (2015) Agenda 2030. Retrieved 24 July 2025 from https://sdgs.un.org/2030agenda

Unteregger R (2025) Project "Pioneer grammar schools in ESD". Future building site, Cudrefin/Switzerland, PH Bern

Wals A, Mathie R (2022a). Whole school approaches to sustainability: exemplary practices from around the world. Wageningen University. Retrieved 24 July 2025 from www.wur.eu/wholeschoolapproach

Wals A, Mathie R (2022b) Reforming education to promote sustainable communities—it takes a whole school. Am Sci 110

Links

Fondation cudrefin02 - chantier de l'avenir (Foundation cudrefin02 - future building site). www.cudrefin02.ch (checked 7/2025)

Foundation for Environmental Education FEE, project Eco-schools. https://www.ecoschools.global/ (checked 7/2025)

Robert Unteregger , Lic. phil., co-founder of the Future Council Foundation (since 1997) and of the Future building site (since 2003) in Switzerland, and lecturer at the Institute for upper secondary education at the Pedagogical University of Berne.

GPSR Compliance

The European Union's (EU) General Product Safety Regulation (GPSR) is a set of rules that requires consumer products to be safe and our obligations to ensure this.

If you have any concerns about our products, you can contact us on ProductSafety@springernature.com

In case Publisher is established outside the EU, the EU authorized representative is:

Springer Nature Customer Service Center GmbH
Europaplatz 3
69115 Heidelberg, Germany

Batch number: 10607214

Printed by Printforce, the Netherlands